Revenge of the Microbes

Revenge of the Microbes

How Bacterial Resistance Is Undermining the Antibiotic Miracle

SECOND EDITION

BRENDA A. WILSON
BRIAN T. HO

Copublication by the American Society for Microbiology and John Wiley & Sons, Inc.

Editorial Correspondence:
ASM Press, 1752 N Street, NW, Washington, DC 20036-2904, USA

Registered Offices:
John Wiley & Sons, Inc., 111 River Street, Hoboken, NJ 07030, USA

For details of our global editorial offices, customer services, and more information about Wiley products, visit us at www.wiley.com.

Wiley also publishes its books in a variety of electronic formats and by print-on-demand.
Some content that appears in standard print versions of this book may not be available in other formats.

Library of Congress Cataloging-in-Publication Data has been applied for

ISBN 9781683670087 (Paperback);
ISBN 9781683670094 (e PDF);
ISBN 9781683673804 (e pub)

Cover Design: Debra Naylor, Naylor Design, Inc
Cover Image: Close up of *E. coli* bacteria. Shutterstock photo ID: 72915925.

Set in 10.5/13pt ArnoPro by Straive, Pondicherry, India
SKY10069529_031224

Dedicated to Abigail A. Salyers and Dixie D. Whitt

whose pioneering voices raised awareness of the once looming antibiotic resistance crisis that is now upon us.

We are indebted to their foresight and are honored to continue their legacy.

Contents

Foreword

MIRACLE DRUGS AND MACHIAVELLIAN BUGS

In 2001 the Amerithrax terrorism attack struck fear in the American public. Postal workers were particularly vulnerable and were very concerned with their safety. Together with her colleague Abigail Salyers, Brenda Wilson became deeply involved in helping the postal workers in her community understand anthrax, from the bacteria, spores, and toxins to the therapeutic use of antibiotics—a crisis course on how to explain important scientific concepts to the public in a clear, comprehensible way. Although anthrax is no longer a national emergency, antibiotic resistance remains a major problem. Written for non-expert readers, *Revenge of the Microbes,* Second Edition clearly explains the problem, provides vivid examples, and explores potential solutions.

Since "the miracle of penicillin" in the early 1940s, antibiotics have had an amazing impact on public health. In addition to saving many people from lethal bacterial infections, antibiotics also opened the door for complex surgeries and medical treatments for cancer and other chronic diseases that would otherwise have resulted in death from secondary infections. However, many pathogens have acquired resistance to effective antibiotics, raising the question of how we can adapt to the growing problem of antibiotic resistance.

The news media is quite effective at spreading the alarm that we are "returning to the pre-antibiotic era," with antibiotic resistance potentially causing over 10 million human deaths per year by 2050. However, evaluating this threat requires an understanding of how antibiotics work, how bacteria develop resistance to antibiotics, and what we can do to counter the threat of antibiotic resistance. This book provides clear answers to each of these questions in a style that is written for the public, avoiding scientific jargon and the scientific minutiae beloved by experts.

The take-home point is that sooner or later bacteria will develop resistance to essentially every antibiotic, whether isolated from nature or designed in a lab. Bacteria are so numerous that even if a mutation that causes resistance is extremely rare, when confronted with an antibiotic that kills sensitive bacteria, the rare resistant mutant can survive and cause disease. Moreover, this resistance can be readily transferred to other bacteria, so widespread resistance can develop rapidly.

Not surprisingly, antibiotic resistance has become a serious global public health problem. The World Health Organization and the U.S. Centers for Disease Control and Prevention have released dire warnings on the global problem of antibiotic resistance. The American Society for Microbiology and other biomedical societies have contributed additional insights about the seriousness of this problem.

Although resistance to existing antibiotics might be mitigated by the development of novel types of antibiotics, due to both technical and economic constraints, there are a limited number of new antimicrobial drugs in the pharmaceutical pipeline. Luckily, new broad-spectrum antibiotics are not the only solution to the antibiotic resistance challenge. Developing rapid diagnostic tests to identify the bacteria causing an infection and its susceptibility to antibiotics will help physicians prescribe the most effective treatment, thereby reducing the problematic use of ineffective antibiotics that promote antibiotic resistance.

Currently it is often difficult to identify pathogens that may persist in sites of the host that are hard to assay, a problem I know as both a microbiologist and a patient. Several years ago, I had a serious antibiotic resistant wound infection that resulted in hospitalization and treatment with a sequence of nine different antibiotics (oral and intravenous) before identifying an antibiotic combination that inhibited the pathogen. However, many people do not have the opportunity (and luck) required to identify an effective antibiotic, and ultimately die from infection.

Ensuring that we have access to life-saving antibiotics relies upon broad public support for the research needed to develop novel solutions to the antibiotic resistance pandemic, and the investment needed to translate this research into potent antimicrobial therapies. Achieving these goals requires that the public understands this problem. *Revenge of the Microbes,* Second Edition provides an interesting, thoughtful overview of antibiotics and antibiotic resistance, addressing each of the challenges in an engaging style that is accessible to public, government, and academic readers. The authors are highly regarded research scientists, teachers, and textbook authors who are passionate about communicating science to the public.

STANLEY MALOY, PHD
Emeritus Professor of Microbiology and Associate Vice President for Research & Innovation, San Diego State University
Editor-in-Chief, Journal of Microbiology & Biology Education
Former President (2005–2006), American Society for Microbiology

Preface

Ever since antibiotic resistance was finally brought to the forefront of public awareness in the 1990s, many discussions of the looming antibiotic resistance crisis (including the first edition of this book) were sheathed in cynical undertones and dripping with disdain. The focus was seemingly always on assigning blame rather than coming up with feasible solutions to the problem. Doctors were derided for lazily overprescribing antibiotics to patients who did not need them. Patients were mocked for demanding these unnecessary prescriptions and then further criticized for not completing their prescribed antibiotic regimens. Farmers were blamed for selfishly and irresponsibly overusing antibiotics for livestock. Pharmaceutical companies were chastised for greedily chasing profits instead of serving the public need. Politicians and policymakers were accused of corruption, or worse, indifference, when faced with outcries for policy changes. All while the public remained generally ignorant and oblivious to the impending antibiotic resistance crisis. Just as with other facets of today's society, all of this mutual antagonism between different groups has fostered unnecessary divisiveness among these communities.

Fortunately, over the past decade, sufficient alarm has been raised that most people acknowledge the importance of the antibiotic resistance crisis and appreciate how the rapidly dwindling list of effective antibiotics is not being replaced by newly developed ones nearly fast enough. While many of the scientific leaders in academic research, industry, and government continue to search for new drug leads, others are doing what they can to respond with stopgaps to lessen the severity of the crisis and to slow its progression until superior solutions can be found.

In the end, the diminishing antibiotic pipeline and the emergence and spread of antibiotic resistance are complex problems affecting people around the globe at all levels of society. Managing it truly is an "everyone" problem requiring multipronged solutions as well as contribution and buy-in from all parties. However, one of the biggest barriers to bringing everyone on board in a concerted effort is the fact that not everyone understands the perspectives, concerns, and motivations of other stakeholders nor do they fully understand what role they can possibly play moving forward. At the core, what is lacking is a clear understanding of the science involved in discovering and developing new antibiotic therapies, as well as the pressures driving the emergence and spread of antibiotic resistance in pathogens.

One of the overarching goals of this new edition is to be a first step forward toward bridging the informational divide by presenting a more holistic view of antibiotics and antibiotic resistance. It is our hope that in reading this book you will gain an appreciation for the challenges associated with bringing a newly discovered compound exhibiting antibiotic activity through the development, testing, and regulatory approval processes to yield a successful and effective new drug on the market, as well as an appreciation for why the pipeline is shrinking. We also hope that you will have a better understanding of how antibiotics work and how resistance can emerge and spread and that you can recognize the many factors that have led to the current antibiotic resistance crisis. Where possible, we have endeavored to include more nuanced discussions of these topics, while acknowledging the perspectives of the different stakeholders. Ultimately, we hope that you will be inspired to lend your support for changes needed to battle the crisis and contribute to possible solutions.

About the Authors

BRENDA ANNE WILSON, PhD, is currently a Professor of Microbiology and Associate Director of Undergraduate Education in the School of Molecular & Cellular Biology, College of Liberal Arts and Sciences; an Inaugural Professor of Biomedical and Translational Sciences in the Carle Illinois College of Medicine; a Professor of Pathobiology in the College of Veterinary Medicine; and Senior Faculty Fellow in the Office of the Vice Chancellor for Research and Innovation at the University of Illinois at Urbana-Champaign, Urbana, Illinois. Dr. Wilson is also a Fellow of the American Academy of Microbiology. She earned her BA degree in biochemistry and German from Barnard College, Columbia University, New York in 1981. She studied as a DAAD graduate Fellow in biochemistry at the Ludwig-Maximilians Universität München in Munich, Germany. Dr. Wilson received her PhD degree in chemistry from the Johns Hopkins University in Baltimore, Maryland, where she received the Ernest M. Marks Achievement Award and an AAUW doctoral fellowship to study the biosynthesis of β-lactam antibiotics in the laboratory of Craig A. Townsend. She then undertook her NIH postdoctoral fellowship training with R. John Collier in the Department of Microbiology and Molecular Genetics at Harvard Medical School, where she began her studies on bacterial protein toxins. Her first tenured faculty appointment was in the Department of Biochemistry at Wright State University School of Medicine in Dayton, Ohio. Dr. Wilson's current research focuses on host-microbe interactions with three main basic science and translational research thrusts. The first is understanding the structure-function, cellular activities, and molecular evolution of bacterial protein toxins and their roles in disease. The second is development of novel post-exposure antitoxin therapeutics and toxin-based therapeutic cargo-delivery platforms. The third is exploiting comparative and functional genomic technologies to explore the role and coevolutionary host-microbe interactions of toxin-producing and extensively drug-resistant bacteria and microbiomes in health and disease.

BRIAN THOMAS HO, PhD, is a Lecturer in Bacteriology at the Institute of Structural and Molecular Biology of the University College London and Birkbeck College, University of London, United Kingdom. He began his research training as an undergraduate researcher in the laboratory of Nancy Kleckner at Harvard University, where he studied changes in chromosome dynamics throughout the bacterial cell cycle. He then went on to earn his PhD degree in the laboratory of John J. Mekalanos at Harvard Medical School, where he studied various aspects of the structure and function of the type 6 secretion system (T6SS) and its effectors in *Vibrio cholerae* and *Pseudomonas aeruginosa*. Continuing as a postdoctoral fellow, his research turned toward studying the T6SS and other contact-dependent secretion systems, such as DNA conjugation, in the context of *in vivo* microbial communities. His current research focuses on understanding how underlying bacterial cell-cell interactions shape larger macroscopic microbial population dynamics and microbial community structure.

1 Antibiotics: What Are They?

WHY IS THERE STILL SOME CONFUSION?

Nowadays, nearly everyone has heard of antibiotics. In fact, most people even have had firsthand experience using them at some point in their life. Whether they were prescribed antibiotics to treat strep throat as a child, or they used an antibiotic cream to manage their teenage acne, or they simply applied an antibiotic ointment to a skin abrasion—the use of antibiotics to treat and prevent infection has become a ubiquitous part of our everyday lives.

However, despite their widespread use, there remains significant confusion regarding several aspects of how antibiotics work and when they should be used. Many patients still ask their doctors: "Why can't I have antibiotics to cure my cold?" or "Why don't antibiotics work against infections like flu or COVID?" And while more informed individuals will know that antibiotics are drugs that specifically target bacteria by inhibiting their growth or outright killing them, these people are then puzzled when doctors sometimes prescribe antibiotics for patients hospitalized with viral infections. What they may not realize is that those doctors are doing so not to directly treat the virus but to prevent secondary infections of bacterial pathogens in individuals with viral-weakened immune systems. Unfortunately, many patients in this situation do not understand why they are taking antibiotics, thinking either that the doctor is just humoring them or, worse, that the antibiotics will actually kill the virus. When these individuals

Revenge of the Microbes: How Bacterial Resistance Is Undermining the Antibiotic Miracle, Second Edition.
Brenda A. Wilson and Brian T. Ho.

subsequently go on social media and spread their incomplete understanding, they sadly end up just eroding the general trust in the medical community.

Further confusion arises when discussing antibiotic resistance. After all, what exactly is antibiotic resistance? Most people correctly recognize that antibiotic resistance means that antibiotics are becoming less effective. However, to many, it is not obvious that antibiotic resistance refers to the bacteria becoming resistant to the drugs rather than individual people becoming resistant to treatment. After all, there are some people who have developed allergies against certain families of antibiotics (e.g., penicillins or macrolides). So in a sense, the misunderstanding is not completely unfounded, as some people's bodies do reject or "resist" these antibiotics. Ultimately, the real challenge is that the two different understandings of the issue have very different implications for what policies, if any, limiting antibiotic usage need to be implemented. If the bacteria are developing resistance, then it makes sense to have antibiotic stewardship policies that limit antibiotic use. However, if resistance occurs only in certain individuals, limiting the use of an antibiotic in one patient will not change the efficacy of that antibiotic when treating someone else.

Perhaps the most problematic source of confusion stems from the dissonance between what people hear from media and what they tangibly experience. Starting in the mid-1990s, stories about antibiotic-resistant bacteria suddenly started appearing everywhere. True to sensationalist form, or perhaps because of a lack of understanding of the problem, the press portrayed what was really a slowly developing, insidious trend to be a galloping crisis. Alarming headlines cautioned against the "Looming antibiotic crisis" and feared the "Return to the pre-antibiotic era on the horizon" and "Superbugs on the march." Writers of such stories conveyed the impression that doom was imminent. Yet 30 years later, antibiotics are still a routine part of our lives. Outside of medical and research circles, one could even be fooled into not knowing there was a problem. A big part of this perception stems from the fact that although many health organizations around the world have declared antimicrobial resistance and the emergence of pan-resistant bacteria ("superbugs") to be an urgent *global* public health threat, thus far antibiotic resistance has disproportionately impacted low- and middle-income countries. However, citizens of the developed world are now realizing that they are no longer safe in their ivory tower, as more recent news headlines make clear: "The superbug that doctors have been dreading just reached the U.S." (*Washington Post*, 2016); "Woman killed by a superbug resistant to every available antibiotic" (*Scientific American*, 2017); "Killer superbugs are coming for you" (*Los Angeles Times*, 2019); and "Drug-resistant infections in hospitals soared during the pandemic" (*New York Times*, 2022). Indeed, American organizations, such as the Centers for Disease Control and Prevention (CDC), the National Institutes of Health, and the American Society for Microbiology, have all joined the World Health Organization (WHO) in undertaking major initiatives to combat antibiotic resistance, including making significant investments in research funding, advocating for policy changes, and creating community outreach and educational programs.

The antibiotic resistance crisis truly is no longer some vague looming threat. We are experiencing it now. The only reason we have been able to keep our heads above water is the collective effort over the last couple decades by academic researchers, industry scientists, and health care workers to mitigate its effects. But unfortunately, we stand now on the brink of losing our ongoing battle against antibiotic resistance. Our ability to discover new antibiotics to replace those lost due to the rapid spread of antimicrobial resistance is simply not keeping up. For example, the rates of resistance to ciprofloxacin, a drug commonly used to treat urinary tract infections, are as high as 93% for *Escherichia coli* and 79% for *Klebsiella pneumoniae* in many parts of the world. *K. pneumoniae* is a major cause of life-threatening hospital-acquired pneumonia and sepsis (bloodstream) infections in newborns and intensive-care unit patients, and it is already resistant to many commonly used antibiotics. Until recently, carbapenem antibiotics were the last-resort antibiotic for treatment of infections by this multidrug-resistant bacterium, but now in some countries even carbapenem antibiotics no longer work in over 50% of patients. Likewise, the WHO reports an estimated half million new cases of infection per year with antibiotic-resistant *Mycobacterium tuberculosis*, of which the vast majority are resistant to the two best anti-tuberculosis drugs. These examples are just the tip of the iceberg. We will revisit these multidrug-resistant pathogens and others in later chapters.

Unsurprisingly, multidrug-resistant infections are extremely deadly. Less than 60% of multidrug-resistant tuberculosis infections are successfully cured. Similarly, patients with methicillin-resistant *Staphylococcus aureus* (MRSA) infections are 64% more likely to die compared to those infected with antibiotic-sensitive bacteria. Patients who are immunocompromised are particularly vulnerable since antibiotic treatment is the only hope of controlling the infections they might acquire. Accordingly, young children and newborns are especially susceptible to antibiotic-resistant bacteria because they do not have fully developed immune systems needed to fight off such infections.

In the constantly changing landscape of scientific and technological advancements, it can be hard to keep up with an ever-evolving understanding of the biological threats impinging on our well-being and the strategies currently available for us to combat them. Misunderstandings due to seemingly conflicting information can be amplified by unfiltered social media to the point that even researchers and medical professionals sometimes get things mixed up. Our goal in writing this book is to help demystify the current understanding of antibiotics and how they work and to resolve some of the confusion surrounding antibiotic resistance.

To this end, let us begin with some foundational information about microbes and a few relevant definitions about antibiotics. In the following chapters, we will try to paint a picture of what life was like before the advent of antibiotics to highlight the impact their introduction had. We will then provide an overview of how antibiotics work, how bacteria have gained resistance, and our current efforts toward combatting this resistance.

WE LIVE IN A MICROBIAL WORLD

Throughout most of Earth's history, life has predominantly been microbial in form. Bacteria were the first life forms to emerge on Earth. They ruled the planet for over 3 billion years before insects, plants, and animals began to appear (Fig. 1.1). During that period, they colonized every part of the world, from the deepest oceans to the highest mountains. They can be found under ice in the Arctic and Antarctic regions. They can be found in boiling hot springs, in nuclear wastes, and as far underground in the land masses as humans have been able to dig. They have even been found in clouds and inside rocks. They survived the volcanic era of Earth's early days, as well as periodic fluctuations in temperatures ranging from ice ages to modern global warming conditions.

Animal- and human-associated microbes comprise an insignificant fraction of the global microbial population. There are roughly 7.5 billion people on Earth, so with an average of about 4×10^{13} associated microbes (mostly bacteria) per adult human, there are roughly 3×10^{23} human-associated microbes. By comparison, estimates place the total number of microbes on Earth at roughly 5×10^{30}—that is about 10 million for every 1 microbe in a person. On top of that, most of the microbes associated with humans are either neutral or benign to people. We live normally with these microbes in and on our bodies without any adverse effects or infections. Taking everything into consideration, only a minuscule portion of microbes actually cause infections; that is to say, only a tiny number of bacteria are pathogens.

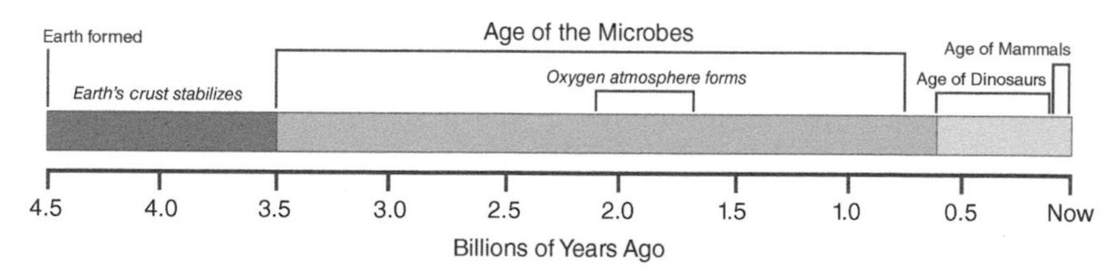

FIGURE 1.1 *The history of life on Earth is dominated by microbial life.*

Our bodies' defenses are remarkably effective in preventing the microbes that we encounter from causing disease. But, alas, nothing is 100% effective. In addition to some microbes having developed ways to circumvent our defenses, breaches in these defenses, such as open wounds, surgery, or cancer chemotherapy, can leave the body temporarily vulnerable to invasion. One of the greatest advances in human health during the past century was the discovery that our natural immune defenses could be augmented with externally applied chemical defenses in the form of antiseptics, disinfectants, and antibiotics.

HOW DO ANTIBIOTICS DIFFER FROM ANTISEPTICS AND DISINFECTANTS?

There are many different chemical compounds that we use to kill or inhibit the growth of microbes, including bacteria, fungi, protozoa (parasites), and viruses. We refer to these compounds with broad killing or inhibiting activity as *antimicrobials.* These chemical agents can be obtained from natural sources (e.g., alcohols, natural acids like vinegar, or toxic metabolites) or they can be artificially produced (e.g., chlorine, iodine, bleach, detergents, and synthetic drugs). Antimicrobials that target only a subset of microbes have more precise names. For example, antimicrobials primarily targeting fungi are called *antifungals;* those that target parasites are *antiparasitics;* those that target viruses are *antivirals;* and those that target bacteria are called *antibacterials.* Although *antibiotic* was originally another word for antimicrobial, for a long time most antibiotics produced commercially or used in medicine specifically targeted only bacteria. Therefore, colloquially, the term *antibiotic* has come to be used interchangeably with *antibacterial.* Accordingly, in this book, we will use the term *antibiotic* to refer to antimicrobials that target bacteria.

So, how do antibiotics differ from disinfectants and antiseptics? Antimicrobial agents that fall into the category of *disinfectants* or *sanitizers* are very harsh chemicals that can only be used externally on inanimate objects or surfaces. These agents prevent infection by reducing the number of microbes that have access to vulnerable areas like cuts or surgical wounds. For example, household bleach, strong acids like hydrogen chloride, and alkylating agents like ethylene oxide are too harsh for human use and therefore are limited to use on nonbiological objects or surfaces. *Antiseptic* or *germicide* is the term used to describe antimicrobial compounds that are less harsh and can be applied to body surfaces. For example, antiseptic hand sanitizers used in hospitals and doctors' offices can be applied to skin. Additional antiseptics are triclosan (an ingredient once commonly used in antibacterial toothpaste, cosmetics, chopping boards, toys, and soaps), dilute hydrogen peroxide solutions, and mercury derivatives such as mercurochrome.

Some compounds, such as iodine and quaternary ammonium compounds (cationic detergents), fall into both categories, depending on the concentrations used and length of exposure.

Antiseptics and disinfectants are generally nonspecific in their killing power and tend to attack multiple targets in microbes. For example, iodine, chlorine, and hydrogen peroxide are strong oxidants that can inactivate many microbial proteins and damage microbial DNA. Similarly, antiseptics that target microbial cell membranes, such as quaternary ammonium compounds, insert themselves into cell membranes, causing cells to leak out vital ions and other small molecules. Unfortunately, these compounds do not act only against microbes—they can damage human cells, too. Therefore, when they are used on humans, their use is limited to topical application in areas where the outer layer of skin consists of dead cells that are unaffected by their toxic activities.

What distinguishes antibiotics from antiseptics and disinfectants is that antibiotics act in ways that are specific for bacteria, and not as general toxins that harm both the bacteria and the human body. The revolutionary feature of antibiotics is that they are designed to be devastating for bacteria but have few if any adverse effects on the human body. Because antibiotics are antimicrobials that are safe to ingest or inject into the body, antibiotics became extremely important for curing and preventing infections. Consequently, they have been touted as the "miracle drugs" of modern medicine. Surgeries and many other invasive medical procedures would not be possible without antibiotics. We will further discuss the vital role of antibiotics in modern medicine in chapter 2.

HOW DO ANTIBIOTICS DIFFER FROM VACCINES?

When we are exposed to bacteria, and especially if those bacteria invade our body, our immune system will attack the bacteria and try to remove them. However, for most disease-causing bacteria, a naïve immune system can take a week or more before sufficient antibodies can be made to mount an adequate immune response. Unfortunately, by that time the invading bacteria will have already caused considerable tissue damage or even death. Using antibiotics can kill the pathogenic bacteria or at least inhibit their growth and replication long enough for the immune system to recognize the invaders and deal with them. The decisive advantage of using an antibiotic is the clearance of the pathogens from the body and a more rapid recovery from the symptoms they caused during the infection, reducing the necessary recovery time from a couple of weeks to a few days.

Vaccination is a related strategy for preventing infectious diseases. It also relies on the immune system for pathogen clearance, but instead of stalling the pathogen until a proper immune response can be mounted, it aims to accelerate the immune response itself. Vaccines work by exposing the immune system to specific structures found on infectious microbes to prime the immune cells to attack those microbes when they are detected in the future. As soon as the priming structures are recognized in the body, the immune cells are stimulated to produce antibodies that facilitate the elimination of the pathogen. This priming enables the full immune response to be mounted within hours rather than days.

Altogether, general cleanliness achieved by using disinfectants and antiseptics and bolstering of the immune system through vaccination are usually sufficient to form an effective bulwark against infectious disease. However, when these preventative measures fail and infection takes hold, because antibiotics are both safe to use in the body and effective during an ongoing infection, we rely upon antibiotics to be the true last line of defense against pathogens.

WHY DON'T MOST ANTIBIOTICS WORK AGAINST FUNGI AND PARASITES?

As mentioned earlier, most antibiotics are antibacterial compounds that have no efficacy against viruses. With a few exceptions, antibacterial compounds are likewise ineffective against fungi and protozoa. As we will see later in chapter 4, bacterial cells are in many ways biologically distinct from animal and human cells. Although many of the molecular components, i.e., proteins, nucleic acids, sugars, lipids, and other metabolites, are functionally equivalent between bacterial cells and human or animal cells, several essential proteins and large molecular structures are sufficiently different that they can be used as targets for antibiotics. Indeed, nearly all antibiotics exploit these differences by targeting the unique features in bacteria, and so they have little to no effect on the functionally equivalent components in nonbacterial cells. In so doing, they can inhibit bacterial growth without causing toxicity toward humans or animals.

Unfortunately, the cellular components of microbes like fungi and protozoa are more like those found in animals and humans than those in bacteria. As such, most antibacterials do not work against them. This similarity to animals and humans is also one of the main reasons we have so few effective antifungals or antiparasitic drugs, and why the drugs we do have tend to cause some degree of toxicity in animals and humans. Additionally, because fungi and protozoa are so similar, the differences that can be exploited frequently are only species- or genus-specific. This narrow spectrum of action means that a drug effective for treating one type of fungal or protozoal disease is often ineffective against other fungi or protozoa. By contrast, because bacteria are sufficiently distinct from humans and animals, the drugs targeting them can be broadly acting against a wide range of bacteria.

WHY DON'T ANTIBIOTICS WORK AGAINST VIRUSES?

In the case of viruses, scientists face a daunting challenge identifying essential viral components to target with antivirals that are not also essential to their human or animal hosts. Viruses, unlike bacteria, fungi, and protozoa, are not free-living microbes. Rather, they are the ultimate freeloaders, relying almost entirely on the cellular machinery of host cells to replicate. Viruses bind to the surface of a mammalian cell and proceed to invade it. During the early stages of this process, the virus sheds its surface covering, called a capsid, to release the DNA or RNA comprising its genetic material into the host cell. This genetic material then highjacks the cell's biosynthetic machinery to reproduce. In most cases, the viral genome and a few virus-associated proteins (the nucleoprotein core) are transported to the nucleus, where the viral genome is copied many times over, and proteins comprising the viral capsid are made at high levels. The newly replicated viral genomes and capsid proteins are then assembled into viral particles and exit the cell, either by bursting out of the cell or by budding out inside vesicles comprising the host cell membrane.

Viral proteins brought in with the viral genome may take part in the viral proliferation process, but most of the action is carried out by enzymes and cell components of the mammalian cell. Therefore, there are very few steps in the viral life cycle that can be targeted by antiviral compounds without also harming the mammalian host cell. These virus-specific targets are, with a few exceptions, limited to the viral surface proteins that mediate attachment to the mammalian cell, the viral proteins that participate in the copying of the viral genome, and those involved in the viral uncoating and budding-out processes. Obviously, these targets are not the same as bacterial cell targets, and so naturally, antibiotics are not effective against viruses.

POINTS TO PONDER

At the end of each chapter, we will include a short "Points to Ponder" section that is intended to highlight some interesting and important questions, issues, or considerations for which there are no easy responses or solutions. They are meant to serve as talking points to spur further discourse regarding the importance of antibiotics, how concerned we should be about antibiotic resistance, and what we should do about the growing rate of antibiotic resistance. With that in mind, what do you think about individuals who reject the notion that antibiotics are essential? These people argue that antibiotics are just weakening the human population by allowing individuals with weaker immune systems to survive and remain in the gene pool. Is this a scientifically sound argument? If so, what societal cost are we willing to accept to pay for such a natural selection-honed immune system?

Keeping with the theme of balancing individual costs and societal benefits is the issue of antibiotic stewardship. As we will discuss later in this book, any time you use an antibiotic there is a small chance that antibiotic resistance can emerge. How important to you personally is preserving the efficacy of our current antibiotics? Are you willing behave in ways that help preserve the efficacy of the antibiotics we currently have—or to put it another way, how much inconvenience are you willing to tolerate personally so that antibiotics will still be available for future generations?

2 The "Miracle" in "Miracle Drugs"

When antibiotics first became available to the general public in the 1940s, they were heralded as "miracle cures" and "wonder drugs." This stands in stark contrast to our current blasé attitude toward them. Our collective consciousness has largely forgotten the integral role antibiotics have played in doubling the average human life span over the last century. To truly appreciate and properly understand why antibiotics were once held in such awe, we need to take a look back to the "before times," when treatments for bacterial infections were not commonplace. Before 1900, not only were bacterial diseases like syphilis, gonorrhea, pneumonia, diarrhea, and tuberculosis (TB) some of the biggest killers in the United States, injuries that would be relatively minor by modern standards were potentially fatal due to the severe risk of bacterial infection. In this chapter, we will review a few examples of past infectious disease scourges and explore some of the present-day medical wonders made possible by antibiotics.

SYPHILIS

If you were sexually active in the pre-20th-century world, syphilis was a very real threat—especially if you or your spouse engaged with multiple sexual partners. Syphilis is caused by a corkscrew-shaped bacterium called *Treponema pallidum* (Fig. 2.1). When left untreated, syphilis progresses through three phases of infection. During the initial phase, or primary infection, growth of the bacteria at the site

Revenge of the Microbes: How Bacterial Resistance Is Undermining the Antibiotic Miracle, Second Edition.
Brenda A. Wilson and Brian T. Ho.

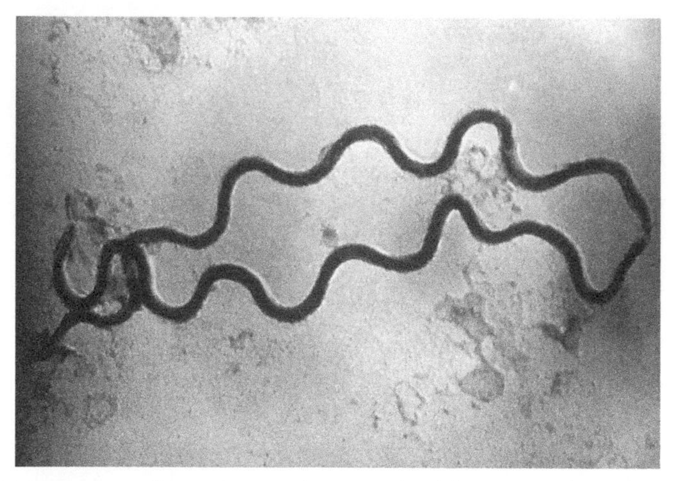

FIGURE 2.1 Treponema pallidum, *the causative agent of syphilis, as seen with an electron microscope. Courtesy of CDC-PHIL (ID# 2392/CDC/Joyce Ayers, 1969).*

of inoculation produces an ugly inflamed sore called a chancre. Although the chancre looks awful, it is seldom painful and can easily go unnoticed. After a week or two, the chancre heals spontaneously, giving a false impression that the infection is gone. The second phase, or secondary infection, occurs several weeks later, when the bacteria enter the bloodstream and can cause several different symptoms to appear. Typically, these include a rash on the palms of the hands or the soles of the feet, as well as fevers, aches, and pains. As in the case of the chancre, these symptoms will spontaneously disappear after a couple of weeks, even though the bacteria are still present. At this point, likely in response to efforts by the body's immune system to combat the bacteria, the *Treponema* bacteria move out of the bloodstream and into bone, liver, muscle, heart, and eventually all other organ tissues of the body, where they are able to lie dormant for months or years. In the third phase, or tertiary infection, the bacteria reactivate. Abscesses and ulcerous masses, referred to as gumma, begin to appear. In some people, the bacteria trigger serious neuromuscular or cardiac (heart) damage, and a painful death can ensue. Because the *Treponema* bacteria can cross the placental barrier, pregnant women infected with syphilis also run the risk of aborting the fetus or giving birth to malformed infants.

During the 15th through 17th centuries, it became clear to physicians, government officials, and historians that syphilis, which was ravaging brothels and bathhouses of the time, was contracted and spread through sexual activity. Despite various determined attempts at implementing regulations and social restrictions, controlling the sexual activities of any population was—as it still is today—a practically impossible task. As such, these efforts ultimately proved futile for eliminating spread of the disease.

In the 16th century, the two main treatments for syphilis were mercury inunction, where the patient was smeared with an ointment containing a high concentration of mercury, and mercury suffumigation, where the patient inhaled or bathed in fumes of mercury, usually while shut in a small hut that was heated to induce sweating. Many patients died from this treatment due to accidental overheating or the toxic effects of mercury. Those who survived often lost teeth and developed sores in their mouths and throats. Despite their sufferings, few patients were fully cured of the disease. The fact that many syphilis sufferers were willing to undergo such drastic and dangerous treatments is indicative of how miserable the disease itself was. Although the early 1900s saw the introduction of more-enlightened therapeutics, such as the arsenic-containing organic compound arsphenamine, they were difficult to prepare, required multiple injections, were fraught with unpleasant side effects, and were not always effective at curing the disease.

Only with the widespread availability of penicillin in the 1940s did syphilis finally become a controllable disease. Not only does penicillin prevent the adverse consequences of the disease from developing, but by eliminating the bacteria from infected individuals, it also helps prevent the spread of the disease among the population. Interestingly, while most other pathogenic bacteria have gradually become resistant to penicillin, *Treponema* is one of the few disease-causing bacteria that has remained susceptible to this antibiotic. We will return to the perplexing question of why some bacteria become resistant to antibiotics so much more readily than others in a later chapter, but for now, just know that thanks to *Treponema*'s continued sensitivity to penicillin, there are some public health officials today who dare to hope that syphilis might one day be eradicated entirely from the developed world. However, despite syphilis being a completely preventable and curable disease, it remains a major challenge for certain high-risk populations around the globe, particularly in low- and middle-income countries, where infrastructural and social barriers hinder access to treatment services.

An example of a more recent, near-tragic case of tertiary syphilis was highlighted in the *New England Journal of Medicine* in 2014 (Chermiak and Silverman, 2014, *N Engl J Med* 371:667), where a woman came to an outreach clinic in rural Uganda with a growing ulcerated mass in the roof of her mouth. Since the mass was initially painless and she had no prior history of syphilis, her previous diagnosis was cancer. She was attempting to sell her home to pay for the surgery to remove the tumor when the mass started to become painful and cause difficulty in speaking and swallowing, prompting her to go to the clinic. Luckily, the outreach clinicians suspected that the mass might not be a tumor but rather a gumma mass formed from tertiary syphilis. When the diagnostic tests came back positive for *T. pallidum*, she was then administered three intramuscular injections of penicillin at 1-week intervals. At a follow-up visit 2 months after the last shot, she was clear of the mass and symptoms. Her life and home were thus saved by quick-thinking doctors and penicillin. This anecdotal case is just a glimpse of not only how potentially devastating syphilis remains today but also how trivial treatment of diseases such as this can become when antibiotics are available.

TUBERCULOSIS

Another dramatic example of the impact of antibiotics is tuberculosis, often just referred to as TB. TB is a lung disease caused by the rod-shaped bacterium *Mycobacterium tuberculosis* (Fig. 2.2). Typical symptoms include a characteristic pallor, persistent coughing, night sweats, and weight loss—this last symptom has led to the disease's historical name, "consumption." In the past, TB was contracted by many well-known artists and writers, who consequently through their works have created an association of beauty and romanticism with the disease. However, what happens inside a TB patient is the very opposite of romantic.

One notable aspect of TB worth mentioning is its means of dissemination. In the case of syphilis, a person could reasonably avoid contracting the disease by living a monogamous lifestyle. Unless you had the misfortune of being raped, having an unfaithful spouse, or being the child of an infected parent, you were generally safe from infection. By contrast, in the case of TB, the disease is acquired by inhaling bacteria introduced into the air by people with active disease. In other words, the main risk factor for acquiring TB is breathing. This

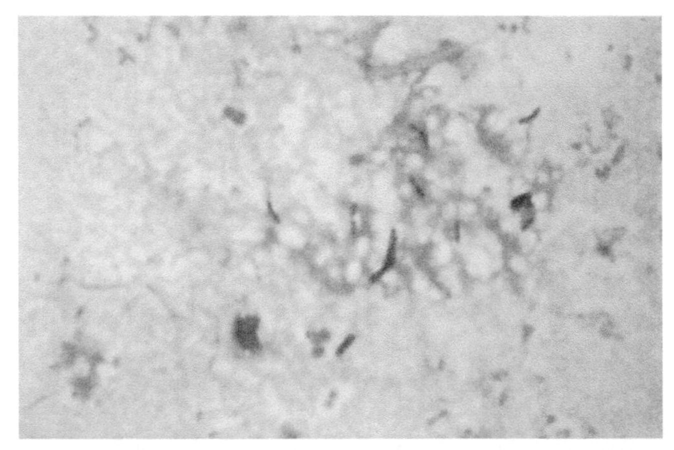

FIGURE 2.2 Mycobacterium tuberculosis, *the causative agent of tuberculosis, as seen in a lung specimen from a patient (dark rod shapes). Courtesy of CDC-PHIL (ID# 4427/CDC/Dr. George P. Kubica, 1979).*

airborne mode of transmission makes it much more difficult to protect oneself from acquiring the disease. Once inhaled into the lung, the bacteria multiply. Most people with a healthy immune system can bring this initial infection under control, but in about 5% of those infected, the bacteria cause so much damage to the lungs that the patient coughs up blood and eventually dies. In some individuals, the bacteria move into the bloodstream and then into various organs, causing an even more aggressive and rapidly fatal disease once known as "galloping consumption."

Today, TB patients take a course of multiple antibiotics to cure their disease. Unlike most antibiotic treatment programs, which are taken for a few days or weeks, anti-TB therapy lasts for at least 6 months and can have unpleasant side effects, ranging from nausea to liver damage. Despite these side effects, antibiotic therapy is still a massive improvement over the slow, agonizing death afflicted individuals once faced. More-fortunate patients of the 19th and early 20th centuries, who could afford the most advanced medical treatments of the time, would abscond away to a sanitarium in the mountains, spending hours in the open air and eating hearty meals in a vain attempt to counteract the wasting away of their bodies as they succumbed to "consumption." As pleasant as surrounding oneself with idyllic scenery might sound, sadly, such treatments were unsuccessful in most cases.

Thomas Mann, the Nobel Prize-winning German novelist, described a sanitarium for TB sufferers in his novel *The Magic Mountain*. The sanitarium is depicted as a delightfully luxurious hotel, but this appealing surface conceals the grim reality of the disease. One of Mann's main characters, a young TB patient called Hans Castorp, arrives at the sanitarium and is taken to his new room by his cousin Joachim, who is also a patient there. At first, Hans is delighted with his room, but Joachim nonchalantly exposes its dark history. "'An American woman died here day before yesterday,' said Joachim. 'Behrens [a sanitarium administrator] told me directly that she would be out before you came. . . . Night before last, she had two first-class hemorrhages, and that was the finish. But she has been gone since yesterday morning, and after they took her away of course they fumigated the room thoroughly with formalin, which is the proper thing to use in such cases.'" Unfortunately, although formalin (an aqueous solution of formaldehyde, a general poison that kills bacteria as well as people) may have prevented acquisition of the bacteria from inanimate objects, it did nothing to prevent transmission through respiratory droplets from other patients.

Less-wealthy people, who could not afford to go to luxurious mountain sanitaria, had their own low-cost versions of the same therapy. In the United States, whole families would live in open wagons, seeking refuge in the fresh air as a "therapeutic" environment. TB patients, called "consumptives," would voluntarily reside in wooden cottages built deep within large caves, such as those in Mammoth

FIGURE 2.3 *Consumptive's Room, Mammoth Cave National Park, Kentucky. Courtesy of Library of Congress Prints and Photographs Division (LC-USZ62-64952).*

Cave National Park in central Kentucky (Fig. 2.3). Others simply perished miserably in their homes or rooming houses.

Perhaps the most inventive, albeit still unsuccessful, strategy for treating TB was devised by rural farmers in the northeastern United States in the 1700s. This strategy reveals the desperation that people of the time felt when they or members of their families faced this dreaded disease. This treatment was based on the hypothesis that TB was caused by vampires. This theory is not as laughable as it might seem at first. Victims of TB often had close relatives who had previously succumbed to the disease and who, after death, might have returned to prey on the living. Also, as the disease progressed, disease sufferers became deathly pale, thin, and ravenous, all classic hallmarks of vampirism according to folkloric belief. The therapy that flowed logically from this understanding of TB was to disinter people who had recently died of the disease and rearrange their bones to make it difficult for them to rise from their graves and walk among the living. Separating the skull from the spinal column and crossing the leg bones across the sternum were common rearrangements. Apparently, vampires were forgetful creatures and did not remember enough anatomy to correct these rearrangements of their body parts. This attempt to treat TB may seem odd or amusing to us today, but in the absence of information about how TB is spread or an effective therapy to treat it, this was the best people could do at the time. Desperation provides a powerful stimulus to the imagination.

Today, TB is still a scourge worldwide and has even experienced a resurgence in several developed countries that have let down their guard. Nearly one-third of the world's population are infected with *M. tuberculosis*, and millions of those infected people develop the full-blown disease and die each year. TB has been a particularly tragic disease in parts of Africa with high rates of HIV infection. HIV debilitates the immune systems of infected individuals, making them more susceptible to TB. Some have called HIV-TB the "one-two" punch of death. Sadly, unlike syphilis, TB is not likely to be eradicated any time soon, even in developed countries, and the few existing anti-TB drugs accordingly remain critically important. More on this topic later in chapter 6.

BACTERIAL PNEUMONIA AND MENINGITIS

As modern-day residents of developed countries, it is easy to relegate syphilis and TB to being mere stories of past centuries, so let us take a look at some diseases that pose a more immediate threat. Pneumonia is a lung disease that was only first successfully treated in the mid-20th century. It is most often caused by the spherical (or coccoid) bacterium *Streptococcus pneumoniae*. As these bacteria grow and divide, they often remain attached to each other, resulting in them being seen in pairs (called diplococci)

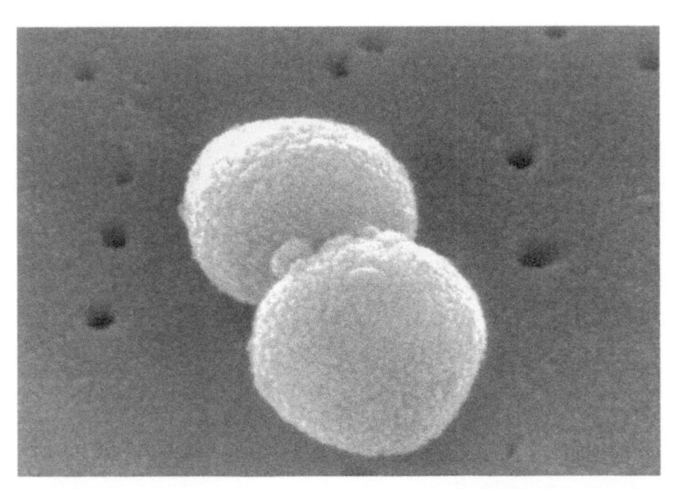

FIGURE 2.4 Streptococcus pneumoniae, *a major cause of bacterial pneumonia and meningitis, as seen with an electron microscope. Courtesy of CDC-PHIL (ID# 263/CDC/Dr. Richard Facklam/Janice Haney Carr).*

under a microscope (Fig. 2.4). Prior to the COVID-19 pandemic, bacterial pneumonia was the leading cause of infectious disease deaths in the United States. Paradoxically, the high incidence of bacterial pneumonia may have been part of what caused it to be overlooked by the news and social media—it is hard to get clicks on a story about a disease that routinely kills tens of thousands of people each year. That many of these deaths are among the elderly adds to our general sense of apathy.

However, right below the elderly, the second-most-common sufferers of bacterial pneumonia are infants. Even the most apathetic might begin to take notice when confronted with a suffering child. *S. pneumoniae* is also the most common cause of earache in children, and although earaches are seldom fatal, they disrupt the lives of many families, especially those with children who undergo recurring episodes. Successful treatment of the infection can be achieved with a 7- to 10-day regimen of oral antibiotics, but oral administration can be challenging for parents and lack of adherence to the regimen is frequently associated with treatment failure or relapse. The deadliest form of *S. pneumoniae* infection in infants is infant meningitis. Before antibiotics, upwards of 90% of children with bacterial meningitis died, and in cases where the disease was not fatal, the survivor was often left blind, deaf, or brain-damaged.

A story that illustrates the dramatic effect of the introduction of antibiotics against *S. pneumoniae* was told in a *New York Times* obituary. In June 1999, Anne Sheafe Miller died at the age of 90, after a long, rich life. Though she was an ordinary citizen, her obituary was nonetheless accorded rare two-column coverage in the *New York Times*. The reason? Anne Sheafe Miller was the first American citizen whose life was saved by penicillin (Fig. 2.5). According to the account, in March 1942, Mrs. Miller was lying near death from an infection following miscarriage in a Connecticut hospital, delirious with chills and fevers spiking as high as 107°F. She had been hospitalized for a month, treated with sulfa drugs (early antibiotics that were the best therapy at the time), blood transfusions, and surgery. Unfortunately, these herculean attempts to bring her infection under control were not working.

In a last-ditch attempt to save Mrs. Miller's life, the doctors injected her with a new drug—penicillin—which was rushed to the hospital from a New Jersey pharmaceutical laboratory. Within 24 hours she was no longer delirious, her temperature had returned to normal, and she was eating normally. As the *New York Times* article put it, "news of Mrs. Miller's full and seemingly miraculous recovery helped inspire the American pharmaceutical industry to begin full-scale production of penicillin." The real news in the obituary, however, was that thanks to penicillin Mrs. Miller survived certain death as a young adult, and she was able to live to the advanced age of 90.

FIGURE 2.5 *Anne Sheafe Miller, pictured here at the age of 75, eventually lived to 90 after penicillin saved her life in 1942 when she was 33 years old and on the brink of dying from a streptococcal infection. Courtesy of the Yale New Haven Hospital Archives.*

SURGERY MADE SAFE

In the pre-antibiotic period, injured soldiers who did not immediately die from the trauma caused by their wounds frequently developed deadly wound infections caused by bacteria such as *Staphylococcus aureus* and *Streptococcus pyogenes*, the latter of which is the primary cause of the so-called "flesh-eating" disease. Amputation was the main treatment when these infections occurred in the arms or legs, but even this drastic intervention had only limited success in saving the patient.

Of the four presidents of the United States assassinated during their presidencies, two of them, Abraham Lincoln and John F. Kennedy, died directly from the damage caused by their bullet wounds. The other two, James A. Garfield and William McKinley Jr., however, might have been saved had antibiotics been available at the time. Both Garfield and McKinley died from infections caused by bacteria introduced into their bodies by the bullet wounds or by the physicians trying to extract the bullets. Garfield was shot twice on July 2, 1881, with one of the bullets grazing his shoulder and the other lodging in his pancreas, where the physicians probed but could not find the bullet. At first, it appeared that Garfield was recovering from his wounds, but he then relapsed with infection-induced fever. His illness progressed over the summer until finally blood poisoning (sepsis caused by bacteria raging throughout his bloodstream) set in, and he died of a ruptured spleen on September 19, 1881. McKinley was shot on September 6, 1901, with one bullet grazing him and the other lodging in his abdomen, where the doctors could not find it. Like Garfield, McKinley at first appeared to be recovering, but then a week later, his health abruptly and rapidly deteriorated. He died from gangrene the next day. Autopsy showed that the second bullet had traveled through the stomach, colon, one of his kidneys, and peritoneum. Although medical practices had improved significantly with the introduction of aseptic techniques in the intervening 20 years between the two assassination incidents, unfortunately for McKinley, these advances could not possibly have prevented the bacteria in the stomach and intestines from being introduced into the tissues and bloodstream.

During World War I, antiseptic solutions, such as diluted bleach, were used to keep wounds free from infection and revolutionized wound care. However, it was not until World War II, and the advent of penicillin, that wound infections would no longer be the primary cause of amputations and death. News of the miraculous incident with Mrs. Miller in 1942, recounted in the previous section, along with other similar stories emerging at the time, led to the Army Medical Corps requesting large-scale production of penicillin—a decision that would save countless lives of wounded servicemen. In fact, there is a famous poster from World War II that depicts a fallen soldier being held by a medic. The caption reads, "Thanks to PENICILLIN . . . he will come home" (Fig. 2.6).

Even in times of peace, wound infections continue to be a serious medical problem in another context, surgery. Surgical procedures have existed for centuries, but only with the discovery of disinfectants, antiseptics, and antibiotics did surgery become the relatively low-risk procedure it is today. Many surgical procedures, e.g., plastic surgery, heart bypass surgery, appendectomy, and cesarean section, are considered routine today, thanks to antibiotics. Unfortunately, the bacteria responsible for postsurgical infections are becoming increasingly resistant to antibiotics, and alarmingly, some appear to be gaining resistance to disinfectants and antiseptics. The emergence of these resistant bacteria means that surgical procedures now bear an increased risk of infection, which in turn means prolonged hospital stays and increased medical costs. But perhaps the most troubling part is that the routine surgical procedures developed over the last 80+ years may ultimately become simply too risky to be performed outside of extreme emergencies.

FIGURE 2.6 *World War II poster extolling the virtues of penicillin. Courtesy of Pfizer Inc.*

CANCER CHEMOTHERAPY AND ORGAN TRANSPLANTS MADE POSSIBLE

Other related areas in which antibiotics have been important adjuncts of medical treatment are cancer chemotherapy and organ transplantation. Most currently available anticancer therapies resemble the old treatments for syphilis, in that they do not directly target the problem (cancer cells), but rather nonspecifically attack all rapidly dividing cells of the body. Unfortunately, a critical portion of these rapidly dividing cells are the immune cells, including white blood cells called neutrophils, which represent the first line of defense against bacterial infections.

Patients receiving aggressive cancer chemotherapy experience a drastic decrease in their white blood cell counts through the course of their treatment. To prevent these patients from succumbing to overwhelming bacterial infections, physicians routinely prescribe antibiotics to be taken while the white blood cell count is low. This preventive therapy tides patients over until their white cell counts return to normal.

Another medical procedure that has been made possible through the advent of antibiotics is organ transplantation. All organ transplant patients receive immunosuppressive drugs, particularly during the 3 months after surgery, to reduce the risk for allergic rejection of the transplanted organ. These immunosuppressive drugs dampen the immune system to the point where infection by overt pathogens as well as opportunistic pathogens (bacteria that normally do not cause disease in healthy people) becomes a major concern. Antibiotics help keep these bacteria in check and are therefore vital for successful outcomes.

IMPROVING QUALITY OF LIFE: ACNE AND ULCERS

Not everyone with an infectious disease ends up dying. Many infectious diseases instead lead to a diminished quality of life. Two notable examples of this phenomenon are acne and ulcers. Acne is thought to be caused primarily by normal skin bacteria, in particular *Propionibacterium acnes,* that get trapped in sebaceous (oil) glands in the skin associated with hair follicles. Sometimes the oily secretions (sebum) dry and plug the end of the gland. The trapped bacteria use the sebum as a nutrient source to grow in the gland. This growth can then cause inflammation and pus (i.e., pimples and zits). Antibiotics such as oral tetracycline or topical clindamycin are often prescribed to treat acne outbreaks. Additionally, acne medications contain antiseptics such as benzoyl peroxides. By limiting the bacterial growth in and on the skin, these treatments can ameliorate some of the symptoms.

For ulcers, the impact of antibiotics is even more clear-cut. For years, physicians were convinced that gastric ulcers were caused by stress. Purportedly, stress led to increased output of stomach acid, resulting in the formation of painful lesions (ulcers) in the lining of the stomach or duodenum. These ulcers could be exacerbated by smoking, eating spicy food, or drinking citrus beverages, alcohol, or caffeine. Treatment usually entailed continuous use of histamine H2 receptor blockers (e.g., cimetidine [Tagamet]) to reduce blood levels of histamine, which results in production of stomach acid. Ulcers that did not respond to these drugs could potentially bleed uncontrollably and would require surgical removal. A big

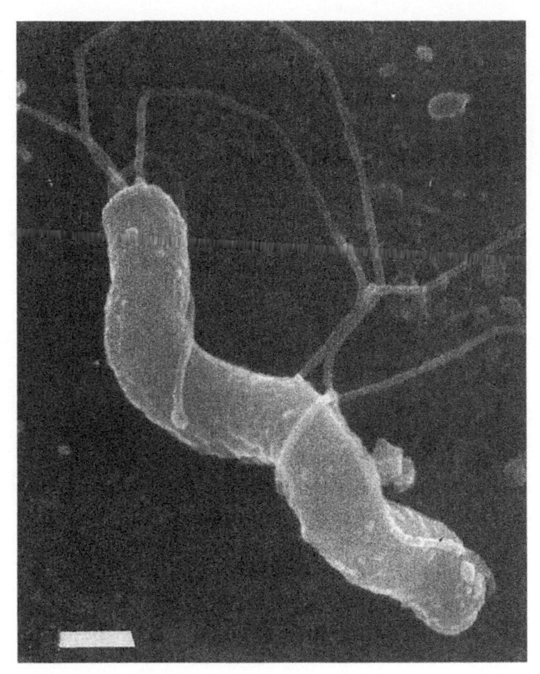

FIGURE 2.7 Helicobacter pylori, *the cause of most gastric ulcers, as seen with an electron microscope. Bar, 0.5 μm. Reprinted from O'Rourke J, Bode G. p 53–67,* in *Mobley HLT et al, ed. Helicobacter pylori:* Physiology and Genetics, *ASM Press, Washington, DC, 2001.*

problem with these treatment options was that they were very expensive, with each costing thousands of dollars and often requiring decades of care. Moreover, patients had to take the drugs continuously, as when they stopped taking them the ulcers would return. However, during the 1980s, a new, seemingly heretical idea began to gain traction: that most gastric ulcers are caused by a spiral-shaped bacterium called *Helicobacter pylori* (Fig. 2.7).

Evidence for the bacterial cause of ulcers had been reported from stomach autopsies of patients with ulcers in the early 1900s, which showed the presence of spiral-shaped bacteria in the stomach pits. For a brief period in the 1940s, New York City hospitals were treating ulcers with the antibiotic tetracycline. However, because no convincing connection could be shown for the presence of bacteria in all cases of ulcers, scientists and doctors began to favor stress-induced acid as the likely cause of ulcer formation. Pharmaceutical companies benefited greatly from the sale of safe and effective drugs like Tagamet that reduced gastric acid and reduced the number of severe ulcers. In 1983, one intrepid scientist, Barry Marshall, took a unique approach to prove that *H. pylori* was responsible for the development of gastric ulcers. He and another volunteer drank a culture of the bacterium. Although Marshall only developed a mild case of gastritis, the other volunteer developed a more severe gastritis that required hospitalization. Interestingly, in both cases, endoscopic examination revealed some inflammation, and subsequent stomach biopsies revealed the presence of the spiral-shaped *H. pylori* in the stomach pits.

Unfortunately, not everyone was convinced by this demonstration. What finally convinced most scientists and doctors were experiments with antibiotic treatment that eliminated the bacteria and thereby cured the ulcers. Finally, after a decade-long struggle, the notion that *H. pylori* was the causative agent of peptic ulcers began to gain acceptance. In February 1994, a National Institutes of Health study panel, comprising many leading experts in the field, endorsed a 2-week course of antibiotics for treating ulcers. Marshall and his mentor, Robin Warren, who first described *H. pylori* in stomach autopsies, received the Nobel Prize in Physiology or Medicine in 2005 for their discovery of the role of *H. pylori* in the pathogenesis of gastritis and peptic (stomach and duodenal) ulcers.

Nowadays, 1-week therapy regimens to clear *H. pylori* are also highly effective, which means that patients who were once spending thousands of dollars every year and yet were still being plagued by ulcers can now be completely cured of all ulcers for only a few hundred dollars (the cost of a short course of antibiotics). Monetary costs aside, despite not being quite as extreme as with syphilis or TB, the quality-of-life improvement afforded by a simple antibiotic treatment was still immeasurable for patients inured to bland diets and years of expensive medications.

In addition to acne and gastric ulcers, there are many other conditions that can now be cured with antibiotics that patients previously would have been forced to endure. For example, infected skin abrasions, urinary tract infections, toothaches, earaches, sore throats, and certain types of debilitating traveler's diarrhea can be reduced in their duration with antibiotic treatment. While many of these infections might be painful, not deadly, reducing the amount and length of suffering clearly improves the quality of life for these patients.

Similarly, several new categories of disease not typically associated with bacteria have recently been shown to potentially have a microbial connection. Examples of such diseases include certain forms of heart disease, preterm birth, neurodegenerative disease (e.g., Alzheimer's), inflammatory bowel disease, and certain cancers. While the precise role of bacteria remains unclear for some of these diseases, the involvement of bacteria opens up the possibility that antibiotics could be used to treat them. These possibilities have driven considerable enthusiasm and interest among some scientists and policymakers toward research in this area.

Ultimately, whether it is curing once-dreaded diseases or enabling new life-saving surgical procedures or even just modestly improving individuals' quality of life, antibiotics have drastically transformed our society. The availability and accessibility of relatively inexpensive life-saving medicines has revolutionized not only what people expect of modern medicine but also the expectations that people have for their own lives. What then, you might ask, would happen if these miracle drugs were suddenly to go away? Unfortunately, this possibility is now becoming the new reality with the emergence of widespread antibiotic resistance in microbes. We will explore this issue in the next chapter.

POINTS TO PONDER

There is a school of historians who claim that antibiotics and vaccines have had only a minimal effect on overall human health. They cite numbers showing that the incidence of infectious disease deaths was already declining significantly well before the introduction of antibiotics and vaccines. No one disputes the fact that changing living conditions, such as reduced crowding, improved hygiene practices, and cleaner water supplies, had a major effect on increased human survival as early as the 16th century. However, from time to time the media have raised the specter of a return to a pre-antibiotic era. Suppose the worst were to happen and all antibiotics were to become ineffective. Would health care return to the level experienced by people living before the 20th century, or are there some ameliorating factors based on what has been learned in the past century, such as improved hygiene and disinfectant use? What changes in medical culture might have to be made to maximize the effects of such ameliorating factors?

3 Bacterial Resistance Threatens the Mighty Reign of Antibiotics

The monumental advances in modern medicine made possible by antibiotics sparked an exciting new era of drug discovery. The race was on to expand our repertoire of antibiotics capable of treating life-threatening infectious diseases. It was quickly discovered that many soil bacteria and fungi naturally produce compounds with antimicrobial properties, which prompted pharmaceutical companies to engage in large-scale screening efforts to find novel lead compounds that could be modified into effective antibiotics. Ultimately, for nearly 30 years after the discovery of penicillin, the market was continuously flooded with new antibiotics. As exciting as these new discoveries were, unfortunately, they ended up masking a growing problem: antibiotic resistance. Essentially, the deluge of new drugs meant that health care providers were able to rely on there always being at least one antibiotic that could treat every infection, despite bacterial resistance to these drugs developing rather quickly. Only after new antibiotic discovery began to slow down did the rosy picture of antibiotics completely solving all our infectious disease woes finally begin to crumble.

ANTIBIOTIC RESISTANCE BECOMES A GROWING PROBLEM

In the years immediately following the advent of antibiotics, academic researchers and some clinicians quickly began to notice that several clinically important bacteria had a high propensity to rapidly develop mechanisms for resisting the antibiotics used to treat them. In some cases, these bacteria would become resistant to a new antibiotic within just a few short years of the drug being first deployed (Table 3.1).

Revenge of the Microbes: How Bacterial Resistance Is Undermining the Antibiotic Miracle, Second Edition.
Brenda A. Wilson and Brian T. Ho.

TABLE 3.1 Evolution of antibiotic resistance

Antibiotic	Year deployed	Year resistance observed
Sulfonamides	1930s	1940s
Penicillin	1943	1946
Streptomycin	1943	1959
Chloramphenicol	1947	1959
Tetracycline	1948	1953
Erythromycin	1952	1988
Vancomycin	1956	1968
Methicillin	1960	1961
Ampicillin	1961	1973
Cephalosporins	1960s	1960s
Carbapenems	1980s–1990s	1980s–1990s
Ciprofloxacin	1987	1990
Linezolid	1999	2003
Daptomycin	2003	2005

Furthermore, antibiotic-resistant strains of several pathogens rapidly developed into the dominant strains of those pathogens circulating among people, particularly in hospital settings. Despite these early signs of danger, so long as the emergence of new resistance did not outpace the development of new drugs, everything would be fine. Unfortunately, after just a couple decades, that is precisely what happened.

As time went on, the rate of novel antibiotic discovery steadily decreased. By the end of the 1980s, the number of new antibiotics entering the marketplace had reduced to barely a trickle. Several compounding technical and economic factors contributed to this slowdown in the drug discovery pipeline. On the technical side, simply identifying novel antibiotic lead compounds is already an extremely difficult undertaking. Finding new leads requires screening libraries of synthetically derived compounds and natural product collections for antibacterial activity. For synthetic compounds, it can be difficult to introduce sufficient molecular diversity into a library, and while natural products often inherently have greater chemical diversity, isolating individual compounds in sufficient quantities from the microbial extracts where they are found can require elaborate procedures. Additionally, regardless of where they come from, simply finding new leads is not enough. Because most of these compounds typically start with very little of the desired antibiotic activity or have other unfavorable properties that make them unsuitable as drugs, further chemical modifications are needed to improve their antibiotic properties.

Altogether, the extensive technical challenges of drug discovery and development mean that the cost of finding new antibiotic candidates is already quite daunting in terms of time, effort, and money. However, this hefty expenditure does not include the enormous monetary and effort costs associated with drug formulation and clinical trials needed to safely bring a candidate to market. Factoring in the high attrition (dropout) rate of candidate drugs that never make it to market, from beginning to end, the entire process of bringing a new antibiotic to market nowadays can take as long as 15–20 years and cost more than $1.5 billion (U.S.). The unfortunate reality is that paying this cost is simply not economically viable for most pharmaceutical companies. Not surprisingly, by the late 1990s, nearly all of the major players in the pharmaceutical industry had shuttered their antibiotic research and discovery divisions.

IMPACT OF MULTIDRUG RESISTANCE (MDR)—WHY DOES IT MATTER?

Despite the diminishing development of new antibiotics, over the decades during which antibiotics were being actively developed, quite a few new antibiotics were added to our arsenal. So, although resistance to just about every available antibiotic could be found in some bacterium in one form or another, it would stand to reason that for any individual bacterial pathogen, at least one of our many other drugs would be effective. Unfortunately, this logic has a couple of major flaws. First, although there are a lot of different antibiotics out there, many of the resistance mechanisms that have evolved in bacteria confer resistance to multiple different antibiotics. Second, not only can bacteria evolve resistance against a given antibiotic, but bacteria can also transfer genes among each other, thereby spreading this resistance horizontally to other bacteria. Lastly, these acquired resistance mechanisms can also accumulate in a single bacterial strain, resulting in the formation of multidrug-resistant (MDR) bacteria. What makes MDR bacteria so scary is that infections caused by these bacteria can potentially become untreatable with any antibiotic, in effect returning us to the pre-antibiotic era.

Fortunately, government health agencies around the world as well as the public at large are finally beginning to take notice. In April 2019, a World Health Organization (WHO) report issued a severe warning to world leaders: "Unless the world acts urgently, antimicrobial resistance will have disastrous impact within a generation." In the press conference announcing this report, Amina Mohammed, the United Nations Deputy Secretary-General and co-chair of the Interagency Coordination Group on Antimicrobial Resistance (IACG), made an urgent call to action: "Antimicrobial resistance is one of the greatest threats we face as a global community. This report reflects the depth and scope of the response needed to curb its rise and protect a century of progress in health. It rightly emphasizes that there is no time to wait, and I urge all stakeholders to act on its recommendations and work urgently to protect our people and planet and secure a sustainable future for all." The report went on to highlight the deficiencies in the current antibiotic development pipeline and listed several priority pathogens urgently needing new treatment options due to the prevalence of multidrug resistance.

MDR BACTERIA—ENTER THE SUPER-SCARY SUPERBUGS!

The emergence of MDR is manifesting epidemiologically in several different ways. Familiar pathogens like *Mycobacterium tuberculosis,* thought to have been dealt with in the developed world, are now making a comeback. Meanwhile, bacteria like *Acinetobacter baumannii,* once only rarely encountered as human pathogens, are now plaguing hospitals around the world. And others normally associated only with animal infections, such as the fish pathogen *Aeromonas hydrophila,* are now infecting humans with devastating consequences. Each of these MDR bacteria is an example of "superbugs" that either are inherently (naturally) resistant or have become resistant (through gain of resistance genes) to many different classes of antibiotics.

Mycobacterium tuberculosis—an old scourge with a dangerous new twist

We already saw the devastation caused by *M. tuberculosis,* the pathogen responsible for TB, in the pre-antibiotic era in chapter 2. In economically well-off countries like the United States and European countries with good health care systems, the incidence of TB infections has been dramatically reduced to near zero levels due to access to effective antibiotics. Unfortunately, in economically challenged countries, extreme poverty and malnutrition have ensured that TB has long remained the world's number one infectious disease killer, only recently being surpassed by the global COVID-19 pandemic. As we will see in later chapters, *M. tuberculosis* is an unusual bacterium with a unique type of waxy coating on its cell surface that is inherently resistant to most antibiotics that are effective against other bacteria. There are only a handful of antibiotics that can be used safely to treat TB, of which the two most powerful are isoniazid and rifampin. Resistance in TB against these two drugs (MDR TB) emerged with the fall of the former Soviet Union and rapidly spread around the world, especially in the Russian Federation, China, and India. MDR TB requires prolonged treatment with second-line drugs that are more expensive, more toxic, and much less effective, with very high rates of treatment failure. In 50% of cases of MDR TB, resistance to these second-line drugs has now also emerged, resulting in so-called extensively drug-resistant (XDR) TB, which currently has a treatment success rate of only 30%. Sadly, in the past 70 years, only two new antibiotics for treatment of MDR TB have been released to the market, making diagnosis and treatment of drug-resistant TB a central focus of the WHO Global TB Programme.

Acinetobacter baumannii—coming to a hospital near you!

After the launch of Operation Enduring Freedom in 2001 and Operation Iraqi Freedom in 2003, a new "superbug" arrived on the scene out of the Middle East, posing a serious global threat. It began with an alarming number of combat-injured soldiers who were being transported to military hospitals, first in Germany and then in the United States. These individuals were succumbing to infection with a highly virulent and pan-drug-resistant bacterium, *A. baumannii*. This common soil bacterium was previously known only as a relatively uncommon opportunistic pathogen, but it was causing deadly bloodstream infections in military service members with normally non-life-threatening blast or burn injuries in Afghanistan, Kuwait, and Iraq. It did not take long for *A. baumannii* to spread from military hospitals to civilian hospitals throughout the United States, Europe, and other countries around the world. The growing global prevalence of this superbug has health officials very worried. The Centers for Disease Control and Prevention (CDC) reported in 2004 that most clinical isolates are MDR, XDR, or even pan-drug-resistant, i.e., resistant to all antibiotics commonly used to treat *A. baumannii* infections, including a group of antibiotics, called carbapenems, that are generally considered to be the "drugs of last resort" for such infections. In 2006, the Infectious Diseases Society of America placed *A. baumannii* on its "Hit List" of top-priority dangerous, drug-resistant microbes in the hope of galvanizing the research community to focus more effort on this pathogen. Today, MDR *A. baumannii* stands second under MDR TB on the WHO's list of critical priority pathogens.

Aeromonas hydrophila—Aimee Copeland's terrifying story

In 2012, doctors from a hospital in Augusta, Georgia, encountered an unusual case of "flesh-eating" bacterial infection in 24-year-old Aimee Copeland, a psychology graduate student at the University of West Georgia (Fig. 3.1). On May 1, Aimee came into the emergency room suffering from a deep cut in her left calf caused by falling from a broken zip line onto rocks in the Little Tallapoosa River. The cut was cleaned and closed with 20 staples, and then she was sent home with antibiotics and pain medication. The next day Aimee returned to the hospital with excruciating pain and a raging infection. Within two days her condition deteriorated to a life-threatening case of necrotizing fasciitis, a "flesh-eating" bacterial infection that destroys tissues, which by day 3 took over her entire leg. While doctors were debriding the bad tissue, she went into cardiac arrest, and after stabilizing her, the doctors made the decision to amputate the leg. Shortly after surgery, she suffered

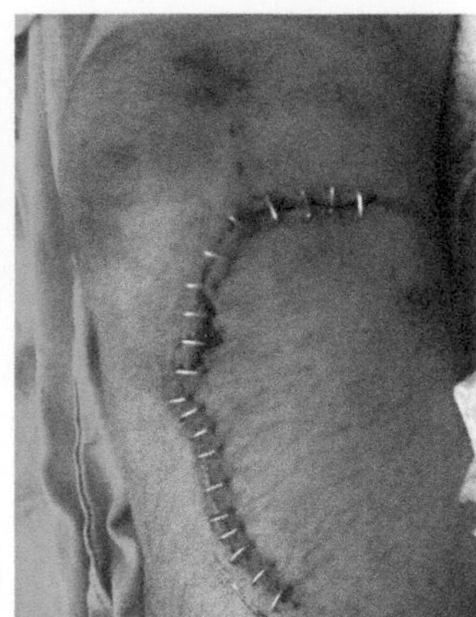
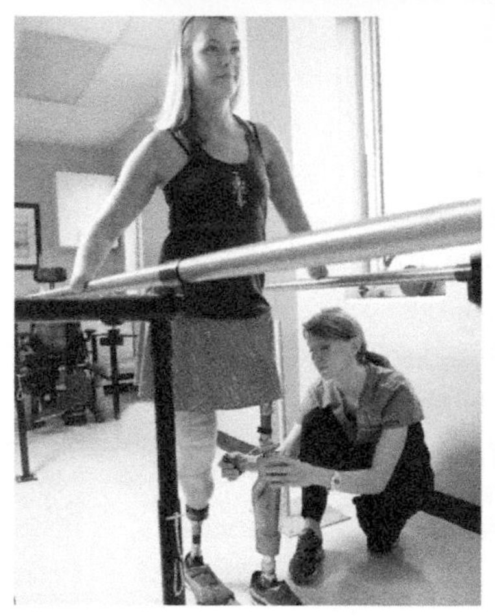

FIGURE 3.1 *Life-altering aftermath of Aimee Copeland's accidental encounter with MDR flesh-eating bacteria. (Top left) Aimee, before her devastating infection with* Aeromonas hydrophila. *(Top right) The laceration on her left calf required 20 staples, but soon became dangerously infected, necessitating amputation. (Bottom left) Necrotizing fasciitis necessitated amputation of her left leg, right foot, and both hands. Aimee is shown here learning how to walk with artificial legs during rehabilitation. (Bottom right) Ten years later, Aimee has adjusted to her new life. Pictured here with her service dog, she is now helping others with disabilities through the Aimee Copeland Foundation. Photos courtesy of Aimee Copeland Mercier.*

renal failure and went into a second cardiac arrest, and following resuscitation she was placed on life support. A few days later, sepsis (uncontrolled bloodstream infection and inflammation throughout her body) caused tissues in her hands and the other foot to die, necessitating further amputations. After weeks of intensive care in the hospital, Aimee slowly recovered and was finally released to go home on

July 2, 2012. After a decade of intensive rehabilitation, she has learned to function well without her limbs and is now a strong advocate for helping others with disabilities.

The clinical diagnostics laboratory isolated and identified this flesh-eating bacterium from Aimee's blood and tissue samples as *A. hydrophila,* a bacterium normally found in most warm freshwater or brackish water environments. While this environmental bacterium is commonly associated with diseases in freshwater fish and amphibians, *A. hydrophila* is not very pathogenic to humans, only occasionally causing self-limiting gastroenteritis from ingestion of contaminated water, and rarely is it reported associated with necrotizing fasciitis. So, what happened in Aimee's case? Why did the antibiotics that the doctors initially prescribed not work? You guessed it! Antibiotic resistance.

It turns out that *A. hydrophila* is naturally highly antibiotic resistant. In recent years, there have been reports of the growing prevalence of resistance to multiple classes of antibiotics in *Aeromonas* strains isolated from storm water drains, lakes, rivers, and other freshwater sources, and especially sewage wastewater. When an environmental MDR bacterium also produces potent toxins that can rapidly damage tissues, you end up with a deadly opportunistic pathogen that can quickly take over a body, with disastrous consequences. Normally, the antibiotics initially prescribed to Aimee should have curbed the spread of the bacteria, but since this particular bacterium was resistant to those antibiotics, the toxins produced by the rapidly growing bacteria, as well as the inflammation caused by the overstimulated immune response, quickly defeated the body's natural defenses, destroying tissues and organs before the infection was brought under control.

A small ray of hope remains . . .

Although there are now some strains of bacteria that are pan-resistant, these strains, luckily, are still in the minority. Most disease-causing bacteria remain susceptible to at least a few antibiotics. There even seem to be some bacteria, such as *Streptococcus pyogenes* (the cause of strep throat and a common cause of wound and bloodstream infections), *Chlamydia trachomatis* (the cause of a sexually transmitted infection that can cause infertility and ectopic pregnancy), and *Treponema pallidum* (the cause of syphilis), that have remained steadfastly susceptible to most antibiotics. Even in the case of bacterial species that are noted for resistance to many antibiotics, there are some strains that have remained susceptible to most antibiotics. Scientists do not know why this is the case—we may be in for some unpleasant surprises in the future if these bacterial slow learners finally catch up with the rest of the class—but for now at least it appears that these serious infectious diseases will remain treatable.

ANTIBIOTIC RESISTANCE ISSUES CROP UP IN UNEXPECTED PLACES—LIKE CROPS

Initially, news stories about antibiotic-resistant bacteria focused on resistance engendered by overuse of antibiotics by physicians. Soon, however, aspects of the antibiotic resistance story connected to agriculture began to surface. The first of these related to concerns about the safety of genetically modified (GM) plants as foods for humans and animals. From there, the antibiotic story grew to include use of antibiotics to treat livestock and antibiotic-resistant bacteria in the food supply. The further discovery that antibiotics could be found in water supplies led to concerns about whether antibiotics could be acting as pollutants or as drivers of antibiotic resistance in environmental bacteria like *Aeromonas* described above.

An ongoing debate in the scientific community often covered in the media centers around the possible hazards posed by antibiotic resistance genes in GM plants. At first glance, it might seem odd to think that the safety of GM plants for human and animal consumption would have anything to do with antibiotic resistance genes, but oddly enough, it does. During the genetic modification process used to construct the earliest GM plants, an antibiotic resistance gene, which was used in the cloning steps performed in bacteria to make the DNA construct, was also introduced into the GM plant along with the gene of interest. For example, Bt cotton, the most widely used GM crop in the United States, is a GM variety of the cotton plant that is resistant to certain insect pests, such as the cotton bollworm. To construct Bt cotton, the Bt insecticidal toxin gene from *Bacillus thuringiensis* (or Bt) that kills the insect was incorporated into the cotton plant genome using a piece of DNA that also contained genes conferring resistance to the antibiotics kanamycin and streptomycin.

Critics of GM plants often raise the question of whether DNA containing antibiotic resistance genes could be transferred from the GM plants to other bacteria in the soil, thereby spreading antibiotic resistance in the environment. They also fear that the DNA might be released from the plant as the plant tissue moves through the human or animal intestinal tract and then, once released, gets transferred into other bacteria that are normally found in the gut. In either case, these potential DNA transfer events could produce new antibiotic-resistant strains of bacteria. Many of the bacteria that normally reside in the intestinal tract of a healthy person can turn deadly if they escape the colon and enter the bloodstream, whether during surgery or some other trauma to the intestinal area. Increased resistance among these intestinal bacteria could increase the risk of not being able to treat such infections effectively. In addition, the intestinal bacteria with these newly acquired antibiotic resistance genes could be released into the environment through sewage systems.

This controversy continues to dominate deliberations of the regulatory commissions assigned to assess possible risks of GM crops, and many of the above-mentioned concerns are often cited in the media and among anti-GM groups. However, with regard to the potential of introducing antibiotic resistance into future GM crops, modern genome-editing technologies, such as TALEN (transcription-activator-like effector nuclease) and the CRISPR (clustered regularly interspaced short palindromic repeat)-Cas system, have now essentially eliminated this concern by enabling precise introduction of the desirable genetic trait (e.g., the Bt insecticidal toxin gene for Bt cotton) without needing to introduce antibiotic resistance genes into the final GM plant product.

HOW'RE YOU GOING TO KEEP BACTERIA DOWN ON THE FARM, AFTER THEY'VE SEEN ANTIBIOTICS?

The old song popular during World War II titled "How're You Going To Keep Them [meaning the soldiers, of course] Down on the Farm, after They've Seen Paree" has an eerie resonance with the current debate about what happens to bacteria that become resistant to antibiotics once they have been exposed to them through agricultural use. Antibiotics are frequently employed in livestock farming to treat and prevent infections and to promote growth of the animals, and the agricultural industry has benefited a great deal from including antibiotics in feedstuffs. However, there is no longer any doubt that this practice has contributed to the antibiotic resistance problem. The low level of antibiotics used in agriculture helps to select for bacteria that can survive in the presence of the antibiotics, i.e., those bacteria that are resistant to the antibiotics.

Why are antibiotics used so much in agriculture?

Antibiotics are used for three purposes in agriculture. The first and by far least controversial is to treat sick animals, but this application only accounts for about 15% of the total antibiotics used in agriculture. The second application, which accounts for 30% of total antibiotic use, is for prophylaxis, the use of drugs to prevent infectious disease. This use is of particular importance for large-scale factory farms where livestock are often housed at high densities. To make matters worse, farm animals bred for maximal production of meat, eggs, or milk typically have weakened immune systems due to their bodies' nutritional resources being diverted away from normal bodily functions in favor of faster growth. This makes prophylactic use of antibiotics essential for preventing infectious diseases from spreading like wildfire through the crowded animal populations.

By far the most controversial agricultural use of antibiotics is as growth promoters (Table 3.2). That is, certain antibiotics seem to give some animals a growth

TABLE 3.2 Examples of antibiotics used in animal agriculture and human medicine

Antibiotic class	Animal use			Human use
	Treatment	Prophylaxis	Growth promotion	
Gentamicin	Yes	Yes	No	Yes
Penicillin	Yes	Yes	Yes	Yes
Cephalosporins	Yes	Yes	No	Yes
Erythromycin	Yes	Yes	Yes	Yes
Fluoroquinolones	Yes	Yes	No	Yes
Sulfonamides	Yes	No	Yes	Yes
Tetracyclines	Yes	Yes	Yes	Yes

advantage. The animals do not necessarily grow larger, but they gain weight more rapidly. Given the slim profit margins most farmers face, even a 5% increase in weight gain (about the effect of the best growth-promoting antibiotic) can be critical. Figures quoted for the percentage of total antibiotics used for growth promotion vary from as low as 15% to as high as 50%. The variation in estimates of growth-promotional use arises from how you define "prophylaxis" versus "growth promotion."

The source of the problem with defining growth promotion is uncertainty about how growth-promoting antibiotics work. Although industry scientists point to published papers on this subject, these papers merely describe changes in the animal's physiology, not the mechanistic basis for these changes. Some scientists have suggested that antibiotics that have been called growth promoters actually work by preventing disease (as prophylaxis), despite the very low levels of antibiotic used for growth promotion. This type of use appears to be more acceptable to the public than growth promotion.

Another possibility is that the growth-promoting antibiotics affect the commensal bacteria living within the farm animals or even directly affect the animals themselves, leading to reduced turnover and excretion of the cells that line the intestine. The constant replacement of intestinal cells is an important defense of the body against bacterial infections because it helps to prevent bacteria that have bound to intestinal cells from invading these cells and moving further into the body. This constant replacement, however, consumes a lot of carbon and energy that could have gone to increase muscle and fat mass. Thus, any treatment that slows the turnover of intestinal cells could significantly increase weight gain.

Arguments about whether an antibiotic is being used prophylactically or as a growth promoter occur because the answer has practical consequences. Can the use of antibiotics as growth promoters be considered nontherapeutic use? The answer to this question depends on one's perspective on how the antibiotics actually promote growth. Some would argue that treatment of a herd or flock in which there are currently no sick animals would fit this use of antibiotics into the category of non-therapeutic use. Others, mindful of the potential for devastating disease in these crowded animal populations, would disagree. If antibiotics promote growth by reducing an animal's load of certain bacteria, then such antibiotics could be considered therapeutic.

Why is agricultural use of antibiotics dangerous?

Complacency about agricultural use of such large quantities of antibiotics persisted for a long time in part because the names of many of the agriculturally used antibiotics were different from those of antibiotics used to treat people. Thus, it appeared to the uninformed person taking a casual interest in agricultural antibiotic use that it had no effect on human medicine. However, many of the agriculturally used antibiotics are closely related in structure to important antibiotics used in humans and can cross-select for resistance to the antibiotics used in treatment of human infections (Table 3.3).

For example, avoparcin, which structurally resembles vancomycin, an antibiotic considered for years to be the drug of last resort for many antibiotic-resistant bacteria like methicillin-resistant *Staphylococcus aureus* (MRSA), was used as a feed additive in Western Europe from 1986 to 1995 (it was never approved for use in the United States). Avoparcin was banned from animal feed in 1997, as soon as it was revealed that this practice led to a sharp rise in the number of Europeans suffering from intestinal infections of vancomycin-resistant *Enterococcus*. Regrettably, as has been found for many other antibiotics, simply halting the use of avoparcin in

TABLE 3.3 Antibiotics used in agriculture that cross-select for antibiotics used in human medicine

Antibiotic used in agriculture	Analogous antibiotic used in human medicine for which resistance is selected	Application for which antibiotic is used in human medicine
Avoparcin	Vancomycin	Postsurgical infections, bacterial pneumonia
Tylosin	Erythromycin	Sexually transmitted diseases, postsurgical infections, lung infections
	Quinupristin/dalfopristin combination therapy	Multidrug-resistant bacterial infections, postsurgical infections, other serious systemic infections
Florfenicol	Chloramphenicol	Second-line usage for antibiotic-resistant bacterial infections

agriculture has not been enough to prevent the spread of vancomycin resistance. It appears that once antibiotic resistance emerges and is given the opportunity to take hold, it becomes impossible for the genes conferring this resistance to disappear.

With this example and many others, the potential danger was becoming abundantly clear, such that by 2002 there were even two bills before the United States Congress that would have drastically limited agricultural use of antibiotics—the Brown bill in the House and the Kennedy bill in the Senate. Unsurprisingly, lobbyists for agricultural pharmaceutical companies and groups representing farmers came out in force to oppose these two bills. Neither bill passed. Luckily, one year later, a force even more powerful than Congress entered the arena on the side of limiting the use of antibiotics as growth promoters in animal husbandry: megacorporation and fast-food giant McDonald's unexpectedly announced that henceforth all its meat suppliers were expected to reduce, then eliminate, the use of antibiotics as growth promoters.

Bad *E. coli*—when Dr. Jekyll becomes Mr. Hyde on steroids . . .

The use of antibiotics places considerable pressure on the microbes in our bodies to withstand the repeated presence of antibiotics. You may have encountered the consequences of this microbial adjustment at some point if you experienced intestinal discomfort or even diarrhea after taking antibiotics. *Escherichia coli* species belong to one group of genetically flexible and highly adaptable bacteria that are part of the normal microbiota residing in human and animal intestines. *E. coli* populations are quite adept at acquiring and exchanging genetic material, carrying genes that encode virulence and/or antibiotic resistance properties, which enable them to survive in the host environment. When a benign (nonpathogenic), resident *E. coli* strain (the "good" Dr. Jekyll) gains a virulence gene that bestows toxic, adhesive, or invasive attributes to the bacterium, the strain can convert into pathogenic, so-called "bad" *E. coli* (the "bad" Mr. Hyde), causing disease in the host, which usually results in diarrhea and subsequent spread through contaminated fecal matter released into the environment or sewage waste systems. When these "bad" *E. coli* subsequently acquire multiple antibiotic resistance genes, they then move into the more dangerous priority pathogen category of MDR/XDR *E. coli* (or Mr. Hyde on steroids).

To illustrate the transformation of *E. coli*, let us consider some recent examples in the evolution of disease-causing *E. coli*. One of the most infamous outbreaks of "bad" *E. coli* occurred in 1992–1993, when undercooked hamburger patties contaminated with a toxin-producing *E. coli* strain with serotype O157:H7 from 73 fast-food restaurants in California, Idaho, Nevada, and Washington infected 732 people, killing 4 children and leaving 178 others with kidney or brain damage. While there had been other, smaller *E. coli* outbreaks from contaminated meat since the early

1980s, the 1993 outbreak brought this particular pathogen to the forefront of public awareness because in this case the *E. coli* strain had acquired a deadly toxin gene that previously was only associated with the bacterium *Shigella,* the causative agent of the bloody diarrheal disease dysentery. Moreover, the Shiga toxin gene associated with the deadly dysentery disease was introduced into the previously nonpathogenic O157:H7 strain through a bacteriophage, which is a virus that infects bacteria. In other words, the toxin gene (like the virus) could continue to spread to other bacterial strains. Fast forward to 2011, when this possibility became reality. Two major outbreaks of *E. coli* in Germany and France were found to be caused by a different *E. coli* strain with serotype O104:H4 that had also acquired the Shiga toxin gene. These two outbreaks ended up causing 54 deaths, 855 cases of bloody diarrhea and kidney failure, and nearly 4,000 cases of acute gastroenteritis.

While the emergence of antibiotic resistance in *E. coli* O157:H7 strains had been relatively slow, what was notable about the O104:H4 strain was that it had broad resistance to multiple antibiotics commonly used to control *E. coli* infections in humans and animals, including tetracyclines, cephalosporins, and fluoroquinolones. For both of these strains, there are increasing reports of difficult-to-treat MDR clinical isolates, presumably due to the favorable conditions within the cow rumen for genetic exchange between *E. coli* strains. Extensive comparison of the genomes of MDR strains of pathogenic *E. coli* from around the world has demonstrated time and again that different *E. coli* strains are able to evolve over time by gaining multiple virulence and antibiotic resistance genes through stepwise acquisition events. Yet we still cannot predict which virulence or resistance genes or in what order those genes will be acquired by a bacterial pathogen to create the next XDR superbug.

ANTIBIOTICS AS A BROADER ENVIRONMENTAL ISSUE—THEY'RE EVERYWHERE!

Why have people become concerned about the agricultural use of antibiotics to the point that even environmental advocacy groups have taken up the issue? Partly, the answer boils down to the amount of antibiotics used. The sheer volume of use in agriculture is enough to get some people worked up. Although the precise figures are controversial, scientists at the CDC and the National Academy of Medicine estimate that as many tons of antibiotics are used in agriculture as are used to treat human disease. Considering that many antibiotics are quite stable in the environment and do not magically disappear after they have been passed through an animal, flushed down the toilet, or leached into groundwater from piles of manure on farms, and you have a situation where tons of antibiotics are ending up in our water supplies, with largely unknown consequences. The worry is that these antibiotics may be driving selection of resistance genes in nonpathogenic, environmental bacteria, which in turn might then transfer their resistance genes to an actual pathogen. However, a more immediate cause for concern is that the antibiotic-resistant bacteria selected on the farm or in a hospital setting can then pass through the food and water supply to enter the human body. If a person acquires a bacterial population in their intestine that contains high concentrations of bacteria resistant to antibiotics, then that person is at greater risk for postsurgical infections that are difficult to treat. The magnitude of this added risk is still up for debate, but few would like to do the experiment of colonizing a substantial portion of the human population with antibiotic-resistant intestinal bacteria to find the answer.

POINTS TO PONDER

Many reporters sensationalize stories about antibiotic-resistant bacteria to grab readers' attention. What would it take to make you read a news article or opinion piece if it was not a sensational scare piece? Economic impact? Specific effects on you or your family?

Antibiotics are not the only chemicals making their way into the water supply. Heart medications and hormones are also being found in surprisingly high concentrations. As scientists are actively investigating the consequences of this type of pollution, many are beginning to suspect that antibiotics and other medicinal compounds could be having significant effects on microbes, plants, and even animals. While the magnitude of this effect remains to be determined, what do you think will be the consequences, if any, of turning water that may be used for drinking or water that flows into nature preserves into a witches' brew of chemicals, including antibiotics?

4 A Brief Look at the History of Antibiotics

When antibiotics were first introduced, they were exalted as miraculous compounds. A dubious mark of medical progress is that most people today have lost this sense of awe. Indeed, as a society, we have come to take antibiotics for granted, often misusing or overusing them without thinking of the potentially disastrous consequences of our actions, such as promotion of antibiotic resistance. To fully comprehend why bacterial resistance to antibiotics is now a major health crisis, it is important to first understand how antibiotics were discovered and came to be a cornerstone of our defense against infectious diseases. So, in this chapter, we will begin with a brief history of their discovery, development, and applications. In the next chapter, we will follow with a brief overview of the different types of antibiotics currently available against bacteria, what their bacterial targets are, and how they work. We will then come back to the problem of how bacteria develop resistance to antibiotics in chapter 6.

PURIFYING SOIL AND UNCOUTH EARWAX

Antimicrobials have not always been as user-friendly as they are now. Today's antibiotics have undergone extensive testing and approval procedures to ensure their safety for human use. As mentioned in the second chapter, the earliest antibacterial compounds, such as mercury and the derivatives of arsenic, were almost as toxic for us as they were for the bacteria. This early approach to therapy was based on what has been called the poison principle. That is, known poisons were

Revenge of the Microbes: How Bacterial Resistance Is Undermining the Antibiotic Miracle, Second Edition.
Brenda A. Wilson and Brian T. Ho.

administered in limited doses in the hope that the bacteria causing the infection would be killed before the person being treated experienced too much toxicity.

Rene Dubos, a research scientist at Rockefeller University, was the first to take a very different view of how antibacterial compounds should work. His view emerged naturally from his lifelong study of soil microbes. Dubos had a high reverence for the purifying properties of soil. In an experiment that was destined to make history, he targeted the bacterium *Streptococcus pneumoniae,* the most common cause of bacterial pneumonia and a major killer for centuries. Dubos' strategy was to mix a laboratory culture of *S. pneumoniae* with an aqueous extract from soil. His theory was that there must be bacteria in the soil that could kill or inhibit the growth of *S. pneumoniae* because the ecological balance could be maintained only if such microbes existed. His idea made sense because, despite its presence in soil and its ability to colonize the human body and cause serious and often deadly disease, *S. pneumoniae* has not taken over the world. Dubos was able to isolate another soil bacterium, *Bacillus brevis,* which produced a substance that was antagonistic to the growth of *S. pneumoniae.*

Unfortunately, the initial form of his anti-*S. pneumoniae* substance had some rather unappealing characteristics that prevented it from gaining popularity as a medicine. As Rollin Hotchkiss, a colleague of Dubos, later described the substance in the book *Launching the Antibiotic Era* (Rockefeller University Press, 1990), the antibacterial compound was a "crude brownish material [that] . . . congealed into a sticky mass as unpleasant as so much uncouth earwax. But it was a powerful wax all right." Ultimately, Hotchkiss and others isolated the active component of this "uncouth earwax" that was able to inhibit the growth of bacteria such as *S. pneumoniae,* a compound we now know as gramicidin. Gramicidin is a small peptide that forms holes in bacterial membranes, which causes the bacterial cell contents to spill out and the cell to die. But, unfortunately, because the membranes of bacteria and humans are very similar in composition, gramicidin proved to be too toxic for internal use in humans, although it is still used as an ingredient in topical antibiotic preparations.

Dubos may not have realized it at the time, but he was on the verge of a new paradigm for fighting bacterial infections and a new era of medicine. The importance of the discovery of gramicidin was that it directed attention to soil microbes as a possible source of antibiotic lead compounds. Subsequently, soil bacteria and fungi proved to be rich sources of antibiotics, such as penicillin and tetracycline, that were much more human-friendly than gramicidin.

THE SULFA DRUGS

Parallel to the quest by Dubos and his colleagues for new natural products of soil microbes, another line of research was to modify existing compounds that kill bacteria to make them less toxic for humans. This approach had been tried with arsenic, but the derivatives of arsenic were still too toxic and were not very effective treatments. As chemists became more sophisticated, they experienced their first significant success: synthetic sulfur-containing compounds called sulfonamides. The discovery of sulfonamides arose from the observation that a red dye called Prontosil could cure some cases of pneumonia. In the early 1930s, scientists identified the active component in Prontosil and determined that it was converted by human cells into an antibacterial compound called sulfanilamide. Sulfanilamide was not nearly as toxic to humans as mercury and arsenic, and thus became the first of what would come to be known as "sulfa drugs."

We now know that sulfanilamide and other sulfa drugs mimic a precursor of the vitamin folic acid (vitamin B_9), called para-aminobenzoic acid, that is important for making genetic material (DNA and RNA) in cells. Bacteria make their own folic acid, whereas humans do not synthesize it themselves and must obtain preformed folic acid from their diet. Because of this difference in metabolism, chemical mimics of para-aminobenzoic acid affect bacteria adversely by inactivating an enzyme the bacteria need to make folic acid, but they have no effect on human cells, which do not need to produce folic acid from such precursors.

A new, but critical principle regarding antibiotics evolved from these two parallel lines of research: the principle of selective toxicity, toxicity against bacteria but not against humans. The best and safest antibiotics are those that act only on bacterial, and not human, targets.

PENICILLIN IS DISCOVERED (ALMOST BY ACCIDENT)

There has been a long-standing debate about who actually discovered penicillin, the antibiotic that, despite the early successes of the sulfa drugs, unquestionably gave the antibiotic revolution its huge momentum. Many historians credit Alexander Fleming, while he was at St. Mary's Hospital Medical School, which is now part of Imperial College London in the United Kingdom, as the discoverer of penicillin. His contribution arose from a series of experiments with a bacterium that was a notorious cause of life-threatening wound infections, *Staphylococcus aureus* (Fig. 4.1). In 1928, Fleming noticed that on some agar plates that had been inoculated with *S. aureus*, which normally forms colonies over most of the surface of the plate, there was an inhibitory zone in which no bacteria grew. This zone had developed around two overlapping colonies of what turned out to be a fungus, *Penicillium notatum*, which was later identified as the producer of penicillin. However, this discovery was not as intentional as textbooks tend to imply.

Fleming was a microbiologist-physician who was interested in a more effective treatment for wound infections. Fleming had focused on *S. aureus* because of its role in war wound infections. He had some interest in new compounds that might be used to control bacterial infections, but his primary focus had been on known

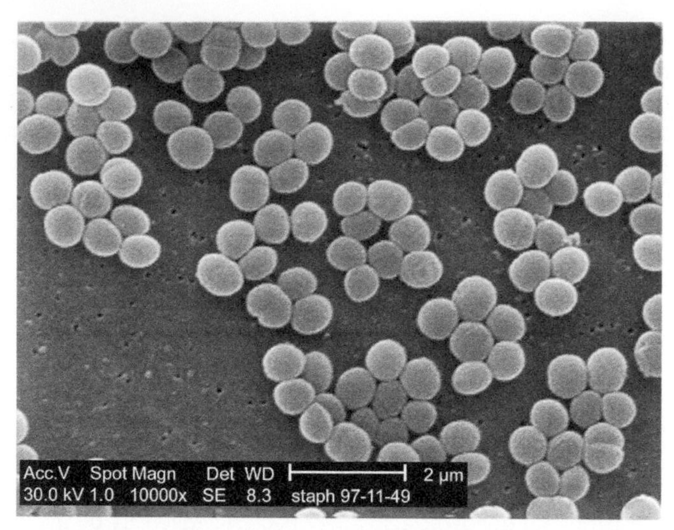

FIGURE 4.1 Staphylococcus aureus, *a cause of serious wound infections, as seen under a scanning electron microscope. Courtesy of CDC-PHIL (ID# 6486/CDC/Matthew J. Arduino, DRPH/Janice Haney Carr).*

antibacterial compounds such as arsphenamine, a derivative of arsenic that had attracted much attention because of its success in curing some cases of syphilis. However, arsphenamine was a rather toxic compound. Fleming knew that some of the bacteria with which he was working could be dangerous. Accordingly, he discarded used agar plates containing colonies of *S. aureus* into trays filled with a disinfectant that was supposed to kill the bacteria. Like some microbiologists of the time, however, he could be careless with his discarded specimens and sometimes let the agar plates pile up too high in the trays before discarding them.

An account of what happened, as written by Norman Heatley, one of the early workers in the area of antibiotic research (*Launching the Antibiotic Era*), highlights the serendipitous nature of how Fleming discovered penicillin:

> In the summer of 1928, Fleming goes on holiday, unaware that he has been chosen by the Fates to take the first steps in introducing the antibiotics to mankind. Having made a wise choice of their agent, the Fates also arranged that one of his plates, inoculated with staphylococci but not incubated, should be contaminated with a spore of *Penicillium notatum*, and that the weather conditions during the subsequent weeks should provide the sequence of rather narrow temperature ranges required to produce the penicillin effect [killing of surrounding bacteria]. Fleming returns from his holiday and goes through the pile of used plates on his bench, looking at them and discarding them into the tray of disinfectant. The plates are numerous and soon they pile up, above

> the disinfectant. But what is this? Gracious heavens, he has discarded the plate! All is not lost, for the Fates have a messenger on hand in the form of Fleming's colleague, D. M. Pryce. Pryce makes his entrance, they chat about staphylococci and to make a point, Fleming picks up some of the discarded plates. The Fates hold their breath. Yes! He picks up *the* plate, looks at it, and says "That's funny . . ." [See what Fleming saw in Fig. 4.2.]. How fortunate that trays rather than buckets were used for discarded cultures and that D. M. Pryce was on hand at the critical moment.

Although Fleming is credited with the discovery of penicillin, some would argue that the scientists who deserve most of the credit for the real successes of penicillin were Howard Florey, a professor at the University of Oxford, and his colleague, Ernst Chain. After the discovery by Fleming, Florey and Chain dedicated themselves during the 1930s and early 1940s to figuring out how to produce, extract, and purify enough penicillin to make the drug available for clinical trials and then for the war efforts. Prior to their discoveries, penicillin was available only in very limited quantities, and most preparations were impure, sometimes containing other toxic contaminants. Without their intervention, penicillin would have had only modest impact. Only when it was produced in large scale and in good purity did penicillin begin to lead to widespread cures of diseases from wound infections to syphilis to bacterial pneumonia. Because of their critical, groundbreaking work, Fleming, Florey, and Chain received the 1945 Nobel Prize in Physiology or Medicine.

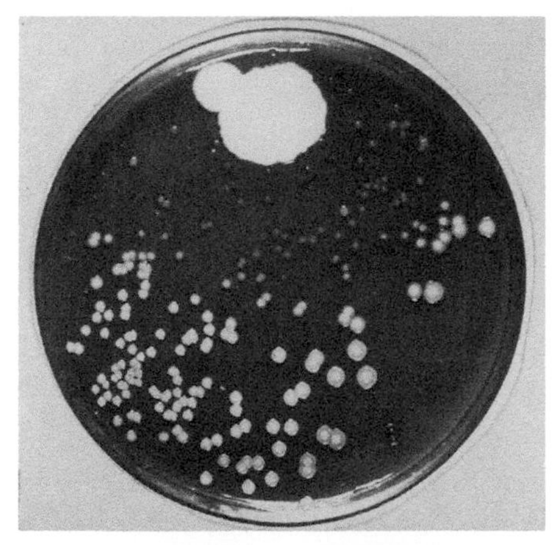

FIGURE 4.2 *Fleming's agar plate of* Staphylococcus aureus *colonies, showing an area of clearing with no bacterial colonies around the much larger overlapping colonies of* Penicillium notatum *at the top of the plate. This zone of clearing (area around the fungus with smaller colonies) indicated that the fungus produced something (penicillin) that was diffusing from the fungal colonies and preventing growth of the bacteria. Credit: Chronicle/Alamy Stock Photo.*

In the race to produce enough penicillin to meet military and civilian demand, the scientists resorted to a variety of growth vessels, ranging from glass bottles that had originally contained popular drinks to bedpans to stackable rectangular ceramic vessels with spouts. The goal was to grow large amounts of the fungus *P. notatum* in order to harvest the culture liquid that contained the antibiotic secreted by the fungus. The main problem was not just the volume needed, but also that culture supernatants from *P. notatum* often had very low antibiotic activity and inconsistent production. Later, Chain realized that the low potency of the culture extracts was because the initial cultures of *P. notatum* were sometimes getting contaminated with a strain of the common bacterium *Escherichia coli*. This particular *E. coli* strain produced an enzyme that degraded penicillin. Thus, even at the earliest stages in the discovery of what would ultimately become one of the most important antibiotic classes ever, the penicillin family of antibiotics, scientists also saw the first evidence that bacteria could become resistant to penicillin. Uh-oh! We will revisit this antibiotic-destroying enzyme and antibiotic resistance in later chapters.

THE RISE OF ANTIBIOTIC DISCOVERY—HOW ANTIBIOTICS BECAME SO IMPORTANT IN OUR BATTLE AGAINST INFECTION

Sulfa drugs and penicillin were the first antibiotics to move from the laboratory into general clinical use. With this encouragement, other soil microbiologists continued the hunt for additional natural products with antimicrobial activity. The 1940s through the 1960s were a golden age of prolific antibiotic discovery. One soil microbiologist, Selman Waksman, received the 1952 Nobel Prize in Physiology or Medicine for his discovery of several new antibiotics, including streptomycin, actinomycin, and neomycin, from his studies with fungi and a large group of soil bacteria called *Streptomyces* (Fig. 4.3). *Streptomyces* species were later found to be a rich source of the majority of the world's naturally occurring antibiotics that are clinically useful. Indeed, many researchers argue that this group of soil bacteria still have untapped reservoirs of new natural products with antimicrobial potential.

Scientists at universities and pharmaceutical companies have since developed many new forms of antibiotics, both naturally occurring and synthetically derived, such that today there are literally hundreds of antibiotics available for treating infections. Unfortunately, the relatively easy access to such a plethora of new and different classes of antibiotics led to excessive use of these antibiotics, not only for medical treatments of infections but also for widespread agricultural and often indiscriminate prophylactic applications. Whenever an antibiotic became ineffective, the pharmaceutical industry responded with renewed efforts toward screening

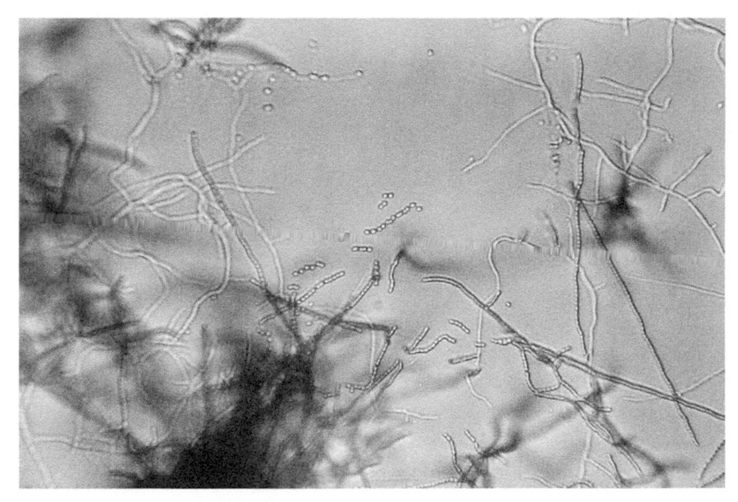

FIGURE 4.3 *A light microscopic image of a culture of* Streptomyces *species grown on tap water agar medium, showing branching bacterial filaments and long chains of bacterial spores. Courtesy of CDC-PHIL (ID# 2983/CDC/Dr. David Berd).*

for more natural products as antibiotic candidates or by using semisynthetic methods to modify the existing antibiotic classes. When these efforts failed to be fruitful, the industry next turned to high-throughput screening of synthetic compounds and powerful robotics to automate the screening process.

Modern genome sequencing technologies also enabled researchers to explore different environments for microbes that have genes for potential antibiotic production. Once researchers began to study the enormous number of microbial genomes coming out of the large-scale microbial genome sequencing efforts, they were surprised to discover that many environmental bacteria and fungi contain many more genes than expected, which could potentially code for the production of new natural products. The implication was that even though these microbes did not make the antibiotic under laboratory conditions, they might make the antibiotic under other environmental conditions. By transferring these genes into artificial systems, the researchers could induce the expression of the biosynthetic enzymes needed to produce the compounds. Once produced, the compounds could be tested for antimicrobial activity. This approach led to a renewed energy in the industry toward uncovering putative novel antibiotics.

Other researchers used genomic technologies in another way to look for new bacterial targets that differed from those of the older antibiotics. The idea is that the researchers could compare the genomes of different microbes for unique features of the microbe that could be used as a target, such as a biosynthetic enzyme involved in a unique microbial process. Once a target protein was identified, it could be isolated and characterized, as well as its structure determined. This target could then be used for screening potential antimicrobials using the large banks of natural products and synthetic compounds. However, as will become evident in later chapters, this quest is not as easy now as it was in the early years of antibiotic discovery, and progress has been slow.

POINTS TO PONDER

Who really deserves the credit for the discovery of penicillin: Fleming or Florey and Chain? This is actually a fairly deep question about assigning credit for a scientific discovery. The credit for such a discovery usually goes to the person or persons who first noticed a phenomenon and performed the first experiments to test the hypothesis the phenomenon suggested. However, effective delivery of such insights to the public is also an important issue to consider. In the case of a work of art, the curators and exhibitors do not get the same degree of credit as the artist who created the work, even though the artist's work would not have received the visibility otherwise. However, in the case of a medical treatment like penicillin, how useful is a drug that remains a curiosity and never makes it to the people who need it? Should those who developed the means for implementing translation from laboratory bench to the patient be given more credit? Clearly the Nobel Prize was shared by all three of these scientists, but there were some who argued for Fleming while others argued for Florey and Chain. What do you think?

5 Bacterial Targets of Antibiotics: How Antibiotics Work

In previous chapters, we have explored the history of our ongoing battle against bacterial pathogens. Doctors and their patients have progressed from being completely at the mercy of bacterial whims to very nearly winning the war with the discovery of antibiotics. But alas, with the emergence of antibiotic resistance, pathogens are now poised to turn the tide against us once again. If we are to have any hope of maintaining our current advantageous position, it is imperative that we fully understand how our antibiotic weapons work, why they are effective, and what vulnerabilities in bacteria they exploit. Ultimately, the winning strategy will likely revolve around identifying key features of the various bacterial pathogens that can be targeted by drugs or other intervention strategies while minimizing the collateral damage to ourselves. In this chapter, we will examine the components of bacterial cells that are typically viewed as good targets for antibiotic development and discuss some examples of how our current antibiotics work against these targets.

KEY COMPONENTS OF BACTERIAL CELLS THAT ANTIBIOTICS TARGET

As discussed in chapter 2, before antibiotics, the primary treatment for infections was simply to give the ailing patient some poisons and hope that their body would survive longer than the pathogen afflicting them. What made antibiotics so powerful that they revolutionized medicine was not their potency against bacteria,

Revenge of the Microbes: How Bacterial Resistance Is Undermining the Antibiotic Miracle, Second Edition.
Brenda A. Wilson and Brian T. Ho.

but rather their specificity for bacteria. By specifically targeting and disrupting key features and processes found exclusively in bacteria, antibiotics are able to inhibit or even destroy bacterial cells without killing the host. Consequently, the very best and safest antibiotics are those that have only bacterial targets and therefore do not cause any toxicity to our cells.

What are some of these differences that can be targeted by antibiotics? Bacteria are prokaryotes. By definition, this means that unlike eukaryotic cells like ours, bacterial cells do not have an internal membrane-encased compartment, called a nucleus, that confines the genetic material (DNA) and the cell's replication machinery. However, there are many other cellular differences. A basic drawing of a bacterial cell is depicted in Fig. 5.1. Bacterial cells are surrounded by a plasma membrane that with the aid of the outer cell wall holds the cellular contents inside as a gelatinous mixture of substances, called a cytoplasm. The cytoplasm contains all of the cell's genetic material as well as most of the proteins and small molecules needed to carry out basic cellular processes, such as DNA replication, transcription (producing messenger RNA molecules called mRNA from the DNA template), translation (producing proteins from mRNA), and metabolism (producing energy from nutrients). Surrounding the cytoplasmic membrane, bacteria also

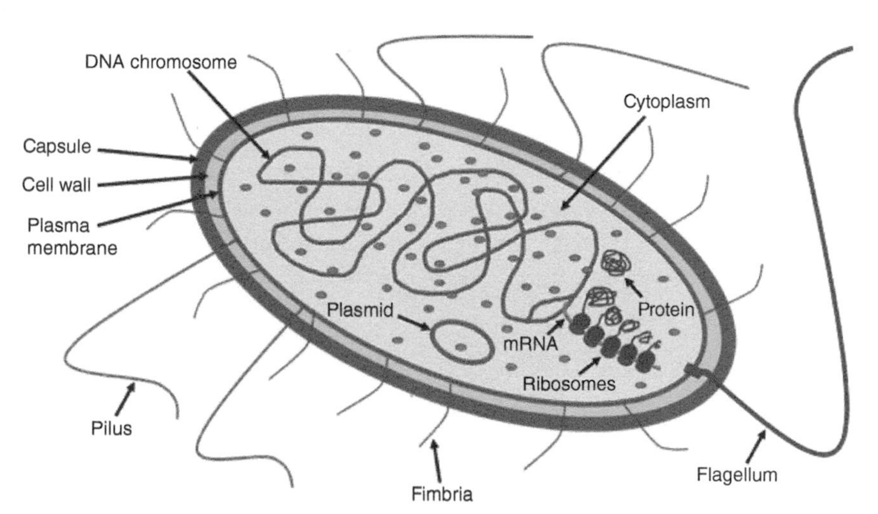

FIGURE 5.1 *Basic structure of a bacterium, showing key components: the outer polysaccharide capsule; the plasma membrane (also called cytoplasmic membrane or inner membrane); the cell wall with peptidoglycan (and an additional outer membrane lipid layer in some bacteria); attachment structures such as longer pili (singular = pilus) and/or shorter fimbriae (singular = fimbria) that mediate adherence to each other or surfaces; motility structures such as a flagellum (plural = flagella); and the cytoplasm, containing various molecules important for cell function such as ribosomes (some attached to mRNA and making proteins), DNA plasmid (often carrying antibiotic resistance genes), and the DNA chromosome.*

have a protective outer structure called a cell wall, which helps to maintain the cell's rigid shape and keep it from bursting (or lysing) under certain environmental conditions, such as extreme salt concentrations or temperatures. Some bacteria also have other outer structures, such as a capsule, pili, or flagella. The capsule is a slimy mesh-like layer of cross-linked sugar molecules that coats the outside of the cell wall and protects the bacterial cell from predation by other microbes, like protists and other parasites, as well as the immune cells of animals and humans. Pili (sometimes called fimbriae when they are short) are hairy appendages comprising polymerized protein subunits that stick out of the cell. The primary function of pili is to enable bacteria to stick to each other or adhere to host and environmental surfaces, but pili are also used for other processes including mediating transfer of genetic material (DNA) between cells, facilitating delivery of toxic proteins from bacteria into eukaryotic cells, and even movement of bacteria across surfaces. Flagella, like pili, are appendages that stick out of cells. However, flagella are usually much larger and can undergo rotational motion. Bacteria use flagella like a propeller to move themselves about in their environment.

To date, the bacterial targets that are most frequently used by the pharmaceutical industry to screen for clinically effective drug candidates are components involved in cell wall biosynthesis, protein synthesis, DNA or RNA synthesis, membrane stability, and folic acid (vitamin B_9) biosynthesis. These targets emerged in early searches for antibiotic activity from natural products, and it was generally straightforward to find chemically modified versions of these compounds that would work as alternative or even better candidates. For decades, this approach was successful in generating many effective antibiotics.

ANTIBIOTICS THAT INHIBIT BACTERIAL CELL WALL SYNTHESIS

Most bacteria are covered by a rigid cell wall that not only defines their shape but also maintains their structural integrity. This rigid coat is necessary because most bacteria live in liquids whose concentration of salts and other small molecules is much lower than the concentration of these same small molecules and others inside the bacterial cell. Since water can flow freely across the bacterial membrane, osmosis is constantly trying to force water into cells, causing them to swell and burst (or lyse). The rigid cell wall helps balance out these osmotic forces by preventing the cell from swelling. Antibiotics that impair the integrity of this cell wall will therefore sensitize bacteria to osmotic pressures, effectively killing them under most physiological conditions.

The rigid bacterial cell wall—essential armor coating for bacteria

The primary component of the bacterial cell wall, which gives it its strength, is a complex mesh-like polymer called peptidoglycan. Among different bacterial species, there are two main types of cell wall. Historically, these types were differentiated based on a specialized staining procedure invented by Hans Christian Gram. Although this Gram stain was developed long before the molecular basis for the difference in the staining was understood, whether a bacterium was positive or negative for the Gram stain was often correlated with how it would behave during infection and how it would respond to different antibiotics. As a result, Gram staining became widely used in clinical laboratories as a simple, early step in the identification and characterization of bacteria in clinical isolates.

In the Gram-staining procedure, bacteria are first attached to a glass slide by heating the slide. Then a solution of two chemicals, iodine and the dye crystal violet, is applied to the bacteria on the slide. The bacteria take up the dye, which stains the bacteria a dark purple-violet color. The bacteria on the slide are then washed gently with an organic solution of acetone and alcohol. This wash removes the dark violet dye from some types of bacteria, rendering them colorless under the microscope, whereas other bacteria retain their dark violet stain. Since the Gram-negative bacteria that lose the violet stain during the acetone-alcohol wash become invisible under a microscope, they must be counterstained with a second, pinkish-red dye, safranin, which gives them a light pinkish-red color. The bacteria that retain the violet dye also stain with safranin, which further darkens the purple-violet color. Bacteria that stain purple-violet are classified as Gram positive, while bacteria that stain pinkish red are classified as Gram negative (Fig. 5.2).

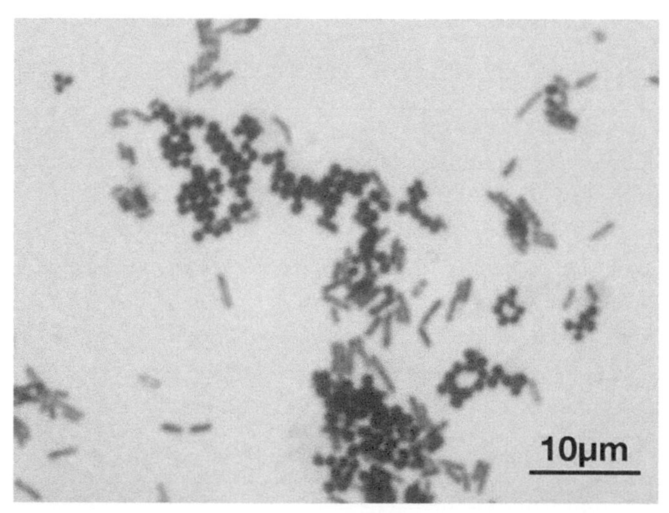

FIGURE 5.2 *A Gram-stained slide of a mixed culture of Gram-positive* Staphylococcus aureus *(purple) and Gram-negative* Escherichia coli *(pink). Courtesy of Y tambe, under license CC BY-SA 3.0.*

We now know that the Gram-positive bacteria stain purple-violet because they have a thick layer of peptidoglycan in their cell walls. Simplified diagrams of the cell walls of common Gram-positive and Gram-negative bacteria, as well as mycobacteria, are shown in Fig. 5.3. Peptidoglycan looks a lot like a chain-link fence. It consists of linear strands of sugars (glycans) that are cross-linked with short chains of amino acids (peptides). Peptidoglycan is present in Gram-negative and Gram-positive cells, but the peptidoglycan layer is much thicker in the Gram-positive cell wall than in the Gram-negative cell wall, and therefore retains the purple dye more. Mycobacteria, which also have a peptidoglycan layer, stain only weakly purple because the unusual waxy mycolic acid-containing outer layer blocks staining.

Bacteria that stain Gram positive have a thick cell wall that consists of many layers of peptidoglycan (Fig. 5.3A). The cytoplasmic membrane of the bacterium is studded with an array of proteins that are exposed on the surface of this membrane. Some of these cytoplasmic membrane proteins join the sugar molecules of the exported peptidoglycan subunits together to make the glycan backbone. Others take the peptide components of the units and join them together with a peptide from an adjacent glycan strand, creating the final cross-linked, mesh-like structure of peptidoglycan. The surface of this peptidoglycan layer is covered with proteins that perform various functions, including allowing the bacterium to bind to human tissues, but they do not provide much of a barrier to small molecules. Antibiotics can diffuse readily through this porous peptidoglycan structure. This is important because the proteins targeted by antibiotics like penicillin and vancomycin are the proteins that assemble the exported sugar-peptide units into the peptidoglycan meshwork. These penicillin- and vancomycin-binding proteins reside in the cytoplasmic membrane or are attached to the membrane surface, and the antibiotics have ready access to them.

Bacteria that stain Gram negative have a more complex cell wall (Fig. 5.3B). Their peptidoglycan covering is only a few layers thick, but it is stabilized and strengthened by a second outer membrane that is unusual. Unlike most biological membranes, this outer membrane does not consist of two layers of phospholipids (called a phospholipid bilayer), but instead consists of a hybrid bilayer with one layer of phospholipids and the other layer of a substance called lipopolysaccharide (LPS). LPS has a lipid component that is embedded in the outer membrane and a long polysaccharide portion that sticks out from the surface of the bacteria. The outer membrane would keep out nutrients and other essential compounds needed by the bacteria if it were not for some open channels (pores) in the outer membrane that allow small molecules to diffuse into the space between the outer membrane and the cytoplasmic membrane. This zone between the inner and outer membranes is called the periplasmic space.

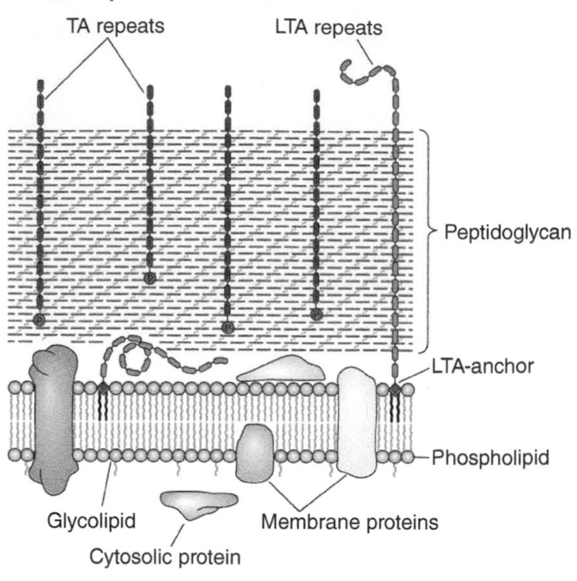

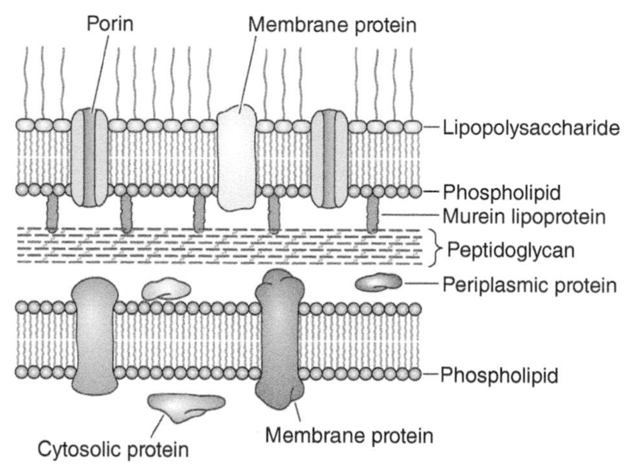

FIGURE 5.3 *Comparison of the cell wall structures of bacteria. (A) The cell walls of Gram-positive bacteria, such as* Bacillus, Streptococcus, *and* Staphylococcus *species, comprise a cytoplasmic membrane with proteins embedded in the lipids and a thick peptidoglycan layer embedded with polymer chains of teichoic acid (TA) links and lipoteichoic acid (LTA) anchored in the cytoplasmic membrane. The thick peptidoglycan layer in Gram-positive bacteria retains the crystal violet dye, and they stain purple. (B) The cell walls of Gram-negative bacteria, such as* Escherichia coli, *also have a cytoplasmic membrane with embedded proteins, but only have a thin peptidoglycan layer. They therefore lose the dye and do not stain purple. Unlike Gram-positive bacteria, the cell walls of Gram-negative bacteria have an additional lipid membrane composed of lipopolysaccharide (LPS) on their outer surface layer that is also embedded with membrane proteins, some of which (called porins) form holes in the membrane. (C) The cell walls of mycobacteria such as the TB-causing* Mycobacterium tuberculosis *have a cytoplasmic membrane embedded with proteins and a thin peptidoglycan layer, but they differ from other Gram-negative and Gram-positive bacteria in that their outer surface membrane-like layer is composed of waxy mycolic acid polymers cross-linked to the peptidoglycan through arabinogalactan sugars and lipoarabinomannan polymers embedded in the membrane and through the outer layer. Mycobacteria also have porin membrane proteins and glycolipids embedded in the outer mycolic acid layer.*

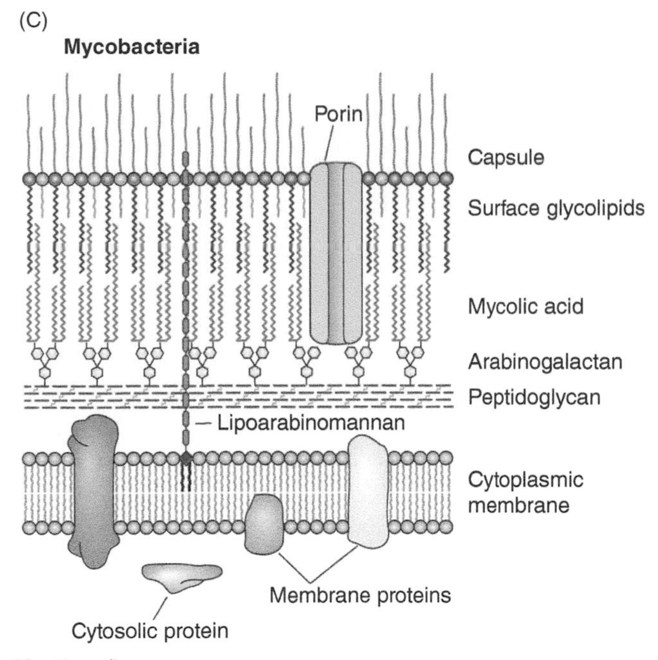

FIGURE 5.3 (Continued)

Mycobacteria have a complex multilayered cell wall that is similar to but also differs from Gram-negative and Gram-positive bacteria (Fig. 5.3C). In addition to their inner cytoplasmic membrane and peptidoglycan layer, they have another unusual arabinogalactan polymer layer, made of linked arabinose and galactose sugars. The arabinogalactan layer is linked to the peptidoglycan and to the characteristic mycobacterial outer membrane comprising a hybrid bilayer with the unusual mycolic acid lipids on the inside layer and a mixture of lipopolysaccharides, including lipomannan, and mycolic acids on the outside layer. Mycobacteria are also coated with an additional thick polysaccharide capsule layer made of arabinomannan (polymers of arabinose linked to mannose sugars) that is anchored to the mycolic acids and to the cytoplasmic membrane through lipoarabinomannan polymers in the outer membrane. This thick waxy mycobacterial cell wall is very difficult for most antibiotics to penetrate, making mycobacteria inherently resistant to most antibiotics that normally work well against Gram-negative and Gram-positive bacteria.

Penicillin and other beta-lactam antibiotics target the bacterial cell wall

Penicillin antibiotics like ampicillin and amoxicillin and the related cephalosporins are truly "cradle-to-the-grave" drugs. Small child with an earache? The doctor will likely prescribe amoxicillin. A young woman with a urinary tract infection?

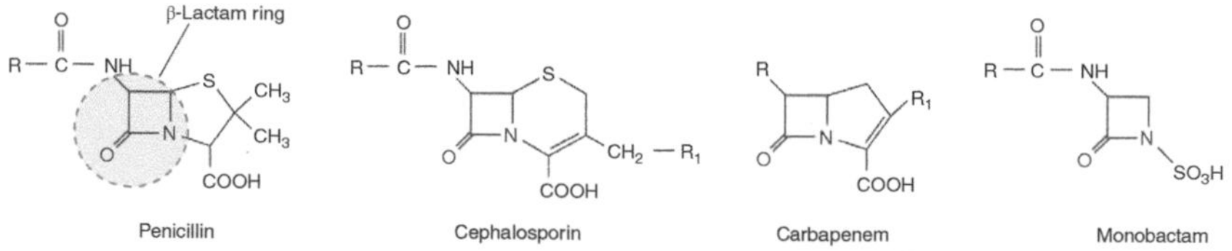

FIGURE 5.4 *Structures of common beta-lactam antibiotics, showing the core four-membered beta-lactam ring (circled) critical for their biological activity in binding to and inhibiting penicillin-binding proteins involved in peptidoglycan biosynthesis. The R group denotes variable components of each antibiotic class.*

Ampicillin is an old standby for treatment. But, what about an elderly man with a case of bacterial pneumonia but who is allergic to penicillin? You guessed it! A penicillin derivative with less risk of allergic response such as cephalosporin or one of the newer monobactams.

The penicillin class of antibiotics has long been a mainstay of antibiotic therapy. The four main members of the penicillin family are shown in Fig. 5.4. Although these molecules look at first as if they are very different in structure, they share a common feature: a four-membered ring called the beta-lactam ring. This ring structure is the active portion of all penicillin family members and is the reason why all members of the penicillin family are called beta-lactam antibiotics. The main motivation for developing the two new members of the beta-lactam antibiotics, carbapenems and monobactams, was to counter increasing resistance of some bacteria to earlier generations of penicillins and cephalosporins and to avoid some allergic responses.

Beta-lactam antibiotics kill bacteria because they undermine the structure of the peptidoglycan layer in the cell wall that serves as a "girdle-like" belt. When the structure is perturbed sufficiently such that it can no longer hold the contents inside, the bacteria explode from the internal pressures, killing the bacteria. The strength of the peptidoglycan meshwork lies in its peptide cross-links. Without those cross-links, the glycan strands are simply wrapped about the bacterial surface like yarn on a ball. As you might know if you have ever toyed with a ball of yarn, the strands of yarn are easily separated from each other, even while they are still wrapped around the ball of yarn. The weakened cell wall that lacks the stabilizing peptide cross-links cannot contain the internal pressure of the bacteria, and the bacteria lyse.

How do the beta-lactam antibiotics undermine the rigid structure of the peptidoglycan layer? The beta-lactam antibiotic binds covalently to the cross-linking enzymes and inactivates them, preventing them from forming the connections between glycan strands and thereby weakening the peptidoglycan meshwork.

Because penicillins bind to these cross-linking enzymes, these bacterial targets have been called penicillin-binding proteins (PBPs). Since the newly synthesized peptidoglycan lacks the strength of the normal meshwork, these antibiotics are most effective when the bacteria are actively growing and the peptidoglycan layer is still relatively thin and less cross-linked.

Other antibiotics targeting the cell wall and how they work

Beta-lactam antibiotics such as penicillin undermine the stability of the very strong peptidoglycan meshwork structure by preventing the peptide cross-links from forming. Other antibiotics, such as fosfomycin (used to treat urinary tract infections) and bacitracin (a common ingredient in over-the-counter antibiotic ointments), act at earlier stages in peptidoglycan synthesis, but they ultimately have the same effect of preventing the peptidoglycan structure from forming correctly.

Peptidoglycan is such a large structure that it cannot be constructed in the bacterial cytoplasm and then exported intact to the bacterial surface through the cytoplasmic membrane. Instead, bacteria must first construct the smaller sugar-peptide building block units of peptidoglycan in the cell's cytoplasm. These sugar-peptide units are then exported through the cytoplasmic membrane and assembled by enzymes on the surface of the cytoplasmic membrane into peptidoglycan, much as bricks can be carried out of a house to construct a wall outside. A special lipid carrier molecule, called bactoprenol, carries out the membrane export process. The antibiotic fosfomycin prevents the synthesis of the peptidoglycan sugar-peptide units, while bacitracin interferes with their export by interfering with the functioning of bactoprenol.

ANTIBIOTICS THAT INHIBIT THE SYNTHESIS OF BACTERIAL PROTEINS

How antibiotics interfere with the bacterial process of protein synthesis

If a teenager goes to a dermatologist for help in controlling acne, the dermatologist is likely to prescribe an oral tetracycline pill, as well as a cream containing the antibiotic clindamycin. Although these two antibiotics are structurally very different and are administered through different routes (orally versus topically), they have the same effect on bacteria. They bind to bacterial ribosomes, the tiny protein factories that translate the genetic information encoded by mRNA into proteins. In doing so, the antibiotics kill the bacteria by preventing them from producing essential proteins.

With one exception (mupirocin), discussed later, all currently available antibiotics that inhibit bacterial protein synthesis do so by binding to the ribosomes, the organelles that string amino acids together into proteins. Bacteria need proteins, just as we do, to perform essential functions and to act as structural components of the cell. Thus, it should not be surprising that the process of bacterial protein synthesis has been a major target for antibiotics.

Bacterial ribosomes, like eukaryotic ribosomes, consist of a large and a small subunit, with each subunit comprising many different proteins and ribosomal RNA (rRNA) molecules (Fig. 5.5). Although a number of these proteins are conserved between bacteria and humans, and despite proteins being made from amino acid building blocks through the same basic mechanism, the components of the bacterial protein-synthesizing machinery are different enough from their eukaryotic counterparts that it is possible to develop chemical compounds that interfere with bacterial protein synthesis without having the same effect in our cells.

During protein production, ribosomes bind to nascent mRNA molecules as they are created from template DNA by RNA polymerase. The ribosome then initiates translation of the genetic code of mRNA into a specified string of amino acids that form a peptide chain (as depicted in Fig. 5.1). To ensure accurate translation of mRNA, every triplet of three consecutive nucleotides, called a codon, corresponds to a single amino acid. This decoding is mediated by another type of RNA molecule, called transfer RNA (tRNA), which brings the specific amino acids to the ribosome

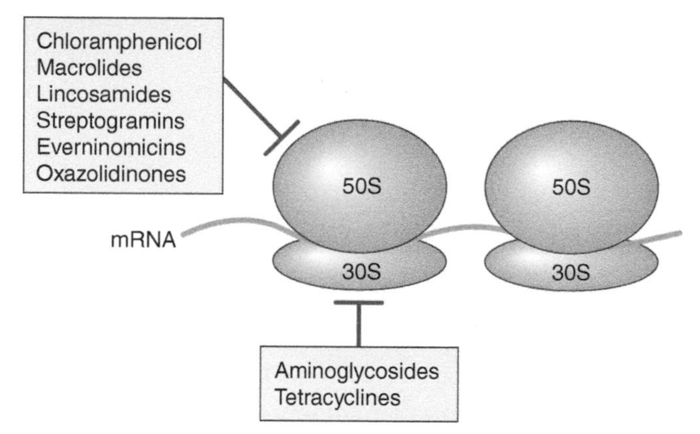

FIGURE 5.5 *The bacterial ribosome is a major target for antibiotics that interfere with protein synthesis. The ribosome comprises two protein-rRNA subunits (a smaller called 30S and a larger called 50S, each made of many molecules of protein and rRNA that form a complex). These two subunits come together to translate the message encoded by mRNA into a protein. The mRNA threads through a groove in between the subunits, and amino acids are delivered to the ribosome by specific amino acid-tRNA molecules that match with the corresponding codons on the mRNA. The ribosome links the amino acids together into the elongating peptide chain, releasing the tRNA carriers, until a stop codon is reached, and the newly synthesized protein is then released from the ribosome. Antibiotics block this process by binding to one of the subunits and preventing their function or by blocking formation of the amino acid-tRNA molecules that serve as precursor building blocks.*

as instructed by the appropriate codon on the mRNA. The ribosome then takes that amino acid and adds it to the growing peptide chain in a process called elongation. The ribosome then moves along the mRNA, translating codons into protein, until it encounters one of the special codon sequences called a stop codon, which signals that the end of the protein has been reached. The final completed protein is then released from the ribosome. The accuracy and order of the amino acids in a protein are typically critical for its structure and proper function and stability. Antibiotics targeting the ribosome can disrupt any of these protein synthesis steps, which in turn can result in various kinds of potentially permanent damage to the bacterial cell.

Streptomycin and other aminoglycosides

As mentioned in chapter 4, streptomycin was one of the earliest antibiotics to enter the market, and it won its discoverer, Selman Waksman, a Nobel Prize. Streptomycin is a member of a family of antibiotics called aminoglycosides, which are among the most widely used antibiotics. This name comes from the fact that the antibiotic is composed of several linked sugars (glycosides) with amino groups attached to them. All members of this antibiotic family function by binding to ribosomes and interfering in some way with the synthesis of bacterial proteins.

The antibiotic streptomycin interferes at the first step in protein synthesis, after the attachment of the amino acids to their corresponding tRNAs and after the two subunits of the ribosome assemble on the mRNA molecule. Streptomycin binds to the small 30S subunit of the ribosome and freezes the ribosome on the mRNA so that it cannot initiate synthesis of the protein. Aminoglycosides other than streptomycin, such as amikacin, kanamycin, and neomycin, prevent later steps in the protein synthesis process, even though they have similar structures. These other aminoglycosides bind to different ribosomal proteins, some in the small 30S subunit and some in the large 50S subunit.

Tetracyclines

Tetracycline, like streptomycin, binds to the small 30S subunit of the bacterial ribosome. The tetracycline family, which includes such antibiotics as doxycycline, oxytetracycline, and demeclocycline, gets its name from its structure, which consists of four fused cyclic rings. In contrast to streptomycin, tetracycline does not freeze the ribosome on the mRNA. Rather, when tetracycline binds to the small 30S subunit of the ribosome, it distorts the structure such that the incoming amino acid-bearing tRNA molecules can no longer interact properly with the mRNA. As a result, the peptide chain is blocked from elongating, and the protein cannot be completed.

Tetracyclines have been on the market since the late 1940s and have been some of the most widely used of all antibiotics. However, in recent years, their popularity has declined because so many bacteria have become resistant to them. When resistance to tetracycline first began to appear, new forms of tetracycline such as doxycycline, minocycline, and oxytetracycline were effective against tetracycline-resistant bacteria. Today, many bacteria have become resistant to all forms of tetracycline. A third-generation tetracycline variety, the glycyl-tetracyclines (such as tigecycline), seems to thwart at least two of the more common resistance mechanisms, so far, and is considered one of the last-resort treatments for resistant bacterial infections.

Despite widespread bacterial resistance to tetracyclines, tetracycline antibiotics are still effective against a few serious diseases. Two examples are anthrax and Lyme disease. In 2001, anthrax-laced letters were sent through the United States postal system. The postal workers were given tetracycline and another type of antibiotic, ciprofloxacin, to prevent inhalation anthrax because the *Bacillus anthracis* bacteria were still susceptible to treatment with these two antibiotics. Likewise, the first-line standard for treating adults suffering from Lyme disease is doxycycline, since the Lyme disease-causing bacterium, *Borrelia burgdorferi,* is still sensitive to this tetracycline derivative. Tetracycline and its derivatives are ideal drugs to use against Lyme disease because of their ability to penetrate deep into tissues, where *Borrelia* organisms are likely to hide out.

Erythromycin and other macrolides

Another widely used class of antibiotics that target the ribosome is the macrolides. Macrolides, such as erythromycin, act by binding to the large 50S subunit of bacterial ribosomes and preventing the elongation of bacterial proteins. These antibiotics have been used for years to treat a variety of bacterial infections, including respiratory infections and wound infections. The macrolides have been particularly useful as an alternative treatment for patients who are allergic to penicillin antibiotics. Another use is the treatment of gastric ulcers. The macrolide clarithromycin was developed to be more acid resistant than erythromycin for oral administration and is now part of the antibiotic therapy used to cure stomach ulcers caused by *Helicobacter pylori,* a bacterium that colonizes stomach pits and causes gastritis and gastric ulcers. Macrolides also have an excellent safety record, with few side effects, mainly mild diarrhea and abdominal pain.

Azithromycin has proven to be a particularly effective macrolide antibiotic as a treatment for sexually transmitted bacterial diseases (STDs), such as gonorrhea, chlamydia, and syphilis. Like clarithromycin, azithromycin was engineered to be more stable to acid, and in addition it could be more readily absorbed by body tissues, such that only one dose needed to be given. As medical staff at STD clinics will tell you, the main concern in cases where patients have bacterial STDs is patient compliance. Clinics have many antibiotics that can cure these infections, but unless the full course of the prescription is taken, these antibiotics may not be able to fully clear infection, which opens the door for recurrence. To make matters worse, the patients who end up in the clinics are more likely to have a drug or alcohol problem or they may be sex workers who may experience repeated exposures to the pathogenic bacteria. Even if told to take even a weeklong course of antibiotics, these patients may stop taking the antibiotic as soon as the symptoms subside but before the bacteria are completely eliminated from the body. Given this context, it should be clear how beneficial it is that only a single dose of azithromycin, administered while the patient is still at the clinic, is needed to successfully treat these infections in most patients. Indeed, with cure rates as high as 97%, further spread of these bacteria can be halted after just one dose of antibiotic.

Clindamycin: an antibiotic class effective against anaerobic bacteria

Another group of antibiotics, the lincosamides, do not resemble erythromycin structurally but have the same mechanism of action of blocking protein synthesis by binding to the large 50S ribosomal subunit. Clindamycin, the antibiotic mentioned earlier in the acne case, is a lincosamide that has been extensively used in the therapy of human diseases caused by anaerobic bacteria.

It was not until the 1980s that clinicians accepted that bacteria that could not replicate in the presence of oxygen (obligate anaerobes) could cause serious human infections. It took a while for them to realize that the human body is not as aerobic as one might think because most of the molecular oxygen in the body is bound to proteins such as hemoglobin. Moreover, damaged tissue that loses its blood supply becomes completely anoxic very rapidly. Obligate anaerobes from the human intestine or the human mouth can very quickly infect such sites if they have the opportunity. The most common cause of this type of infection is a major bacterium found in the human intestine, called *Bacteroides*. Such infections, which can occur after abdominal trauma resulting from surgery or accidents, can be deadly because *Bacteroides* is naturally resistant to antibiotics such as aminoglycosides and many strains are widely resistant to many other antibiotics, including the tetracyclines and penicillins. Clindamycin burst on the infectious disease scene as the antibiotic that solves the problem of infections caused by these anaerobic bacteria.

Mupirocin: an antibiotic that targets amino acid-tRNA synthesis

Each type of amino acid is attached to its specific partner tRNA to form the different amino acid-tRNAs by enzymes called aminoacyl-tRNA synthetases. These enzymes are the target of an antibiotic called mupirocin, which inhibits their activity. This antibiotic target, which is not a part of the ribosome, is the exception to the rule that protein synthesis-inhibiting antibiotics bind to the ribosome. Mupirocin was ignored for years because it is too toxic for internal use. It has been used primarily as a topical antibiotic to eliminate antibiotic-resistant *Staphylococcus aureus* from the noses of hospital workers, who could potentially transmit these bacteria to vulnerable patients. In recent years with the rise of antibiotic-resistant *S. aureus*, pharmaceutical companies have revisited mupirocin as a lead for new classes of antibiotics that interfere with protein synthesis through inhibition of tRNA synthetases.

ANTIBIOTICS THAT TARGET BACTERIAL DNA AND RNA

Antimicrobials that target DNA and RNA can do so by chemically modifying the DNA to generate breaks in the double-helix strands, binding to DNA and preventing replication of the DNA or transcription of the DNA into RNA, or binding to and inhibiting enzymes that are involved in DNA replication or transcription processes (Fig. 5.6). Alternatively, antimicrobials could inhibit the biosynthesis of the nucleotide building block precursors. These antimicrobials are useful clinically only if they interfere preferentially with bacterial and not mammalian processes or enzymes. Consequently, some antimicrobials that directly modify DNA, such as mitomycins that cross-link DNA, or insert into DNA and induce double-strand breaks, such as bleomycins and actinomycins, act on both microbial and human DNA and therefore are far too toxic to be used as antibiotics. We will not consider those here, and instead, will cover a few examples that have been shown to be relatively safe and effective antibiotics because they selectively act on bacterial processes.

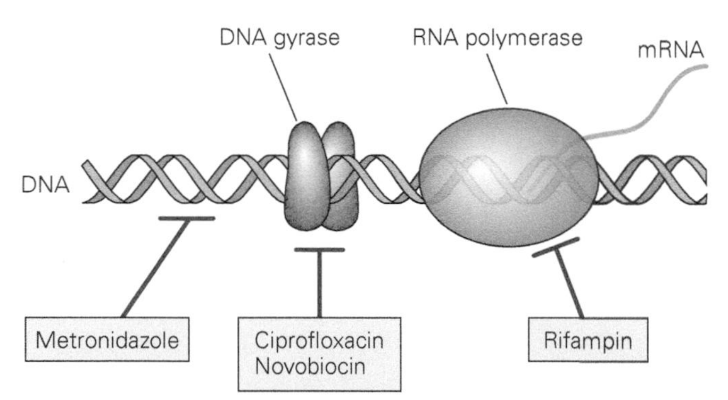

FIGURE 5.6 *Antibiotics that interfere with DNA and RNA synthesis. Key bacterial-specific targets include the bacterial enzyme DNA gyrase, which is an important protein that unwinds the DNA supercoils that form during DNA replication; and the bacterial enzyme DNA-dependent RNA polymerase, which is involved in transcription of DNA into mRNA. Metronidazole is a DNA-damaging antibiotic that works specifically in bacteria and not mammals because only bacteria have the enzymes needed to convert the drug into its active form.*

Metronidazole—bacterial-specific DNA damage

Metronidazole has become a mainstay antibiotic for treating anaerobic bacterial infections, such as those caused by the anaerobic pathogen *Clostridioides difficile* (formerly called *Clostridium difficile*), and especially those that are resistant to other antibiotics. Metronidazole causes breaks in the double-stranded DNA helix; in other words, this antibiotic acts like a mutagen to damage DNA. But in order to do this damage, metronidazole must first be activated before it can attack. This activation is done by specific reducing enzymes, called ferredoxins and flavodoxins, that are found only in anaerobic microbes, ones that can live only in the presence of little or no oxygen. Aerobic microbes, animals, and humans do not have these metronidazole-activating enzymes, and so are not affected by the DNA-damaging properties of the antibiotic that only anaerobic microbes can experience.

Ciprofloxacin and other fluoroquinolones—targeting DNA replication

Fluoroquinolones, such as ciprofloxacin (Cipro), have become widely used front-line antibiotics. A close relative, the original quinolone antibiotic, nalidixic acid, has been around since the 1960s. Although nalidixic acid was used successfully for treating urinary tract infections, such high doses were needed that toxic side effects were common. Simply adding a fluorine atom to the basic quinolone structure made these antibiotics much more effective against bacteria, so that much lower concentrations could be used for therapy. Due to the lower concentration needed for efficacy, side effects became less common, and so the use of fluoroquinolones increased, particularly for problematic antibiotic-resistant bacterial infections.

An earlier statement that fluoroquinolones inhibit DNA replication might seem to suggest that these antibiotics inhibit the copying of a DNA strand during replication, but the mechanism of action of the fluoroquinolones is a bit more complicated than that. As a bacterium elongates prior to dividing to form two daughter cells, its chromosome is copied by DNA polymerase so there will be two chromosomes, one for each of the daughter cells. In order for DNA polymerase to gain access to the DNA it is going to copy, the tightly wound strands of DNA have to be unwound. After the copying is completed, the DNA of the two resulting copies of the chromosome must be wound up again. This DNA winding, called supercoiling, is so tight that the DNA double helices collapse into a knot-like structure. Such tight winding is necessary to make the long DNA strands compact enough to fit inside the cell.

Fluoroquinolones bind to proteins involved in the winding and unwinding of DNA during replication. DNA gyrase is an enzyme that is involved in returning newly replicated DNA into its supercoiled form. Fluoroquinolones act by binding tightly to DNA gyrase and distorting its structure sufficiently to prevent the gyrase from doing its job. Human cells have enzymes with similar functions, but they are different enough from the bacterial DNA gyrases that the human cell enzymes do not bind fluoroquinolones and are thus not affected by the antibiotic. Another target of the fluoroquinolones is an enzyme called topoisomerase IV. This enzyme, like DNA gyrase, also plays a role in the supercoiling process by helping to separate the daughter chromosomes during the final stage of DNA replication so the DNA chromosomes can then be compacted into the newly divided bacterial progeny.

Rifampin—targeting RNA synthesis

Rifampin is a semisynthetic antibiotic of the rifamycin class produced by the soil bacterium *Amycolatopsis rifamycinica* that is used to treat mycobacterial infections, Legionnaires' disease, *Haemophilus influenzae,* meningococcal disease, and some infections where tetracyclines are unable to be used. Rifampin blocks the transcription of DNA into mRNA by binding to the exit site of the enzyme DNA-dependent RNA polymerase, blocking the newly synthesized mRNA molecules from leaving the protein and thereby preventing mRNA from being translated into protein.

Sulfonamides and trimethoprim—targeting folic acid (vitamin B_9) biosynthesis

A commonly prescribed antibiotic combination is sulfanilamide (a sulfonamide derivative) and trimethoprim, sometimes referred to as sulfa-trimethoprim. These antibiotics interfere with the production of the essential vitamin B_9, tetrahydrofolic acid (THF), which is essential for the growth of bacteria. The role of THF and its derivatives in metabolism is to serve as a conduit for transporting single-carbon groups (such as methyl groups) from one biological molecule to another during biosynthetic processes. It is essential for making nucleic acid building block precursors, purines and pyrimidines; for making the amino acid methionine from its precursors; and various other metabolic reactions in cells.

Sulfanilamide and trimethoprim work by mimicking the substrates of two different enzymes in the pathway of THF biosynthesis. In this case, the enzymes try to carry out their normal reactions but mistakenly use the antibiotic instead of the true substrate for the reaction. The antibiotic inhibits the activity of the enzyme so that instead of smooth and efficient synthesis of THF, the reaction stops.

Most bacteria make their own THF because they lack the transport molecules needed to take it up into the cell. This makes the bacterial THF biosynthetic enzymes good targets for developing antibiotics. Indeed, as mentioned in chapter 4, sulfanilamides (the so-called sulfa drugs) were some of the first antibiotics that became available in the 1930s. We humans are equally dependent on folic acid derivatives, but the sulfa-trimethoprim combination is not toxic for us because we get these folic acid compounds from our diet rather than synthesizing them ourselves.

Unfortunately, resistance to sulfa drugs or trimethoprim develops easily if these antibiotics are taken separately. The protein target of the antibiotic mutates so that it still works but no longer binds the antibiotic. The solution to preventing development of resistance, as well as to increasing the efficacy of the preparation, was to combine sulfanilamide with trimethoprim. Because the two compounds inhibit different enzymes involved in THF biosynthesis, the probability of a bacterium becoming simultaneously resistant to both antibiotics was far lower than the probability of a mutation leading to resistance to just one of them. A caveat is that this strategy only works if the antibiotic combination is applied simultaneously. If a patient first takes one antibiotic and then takes the second one, all bets are off, because the probability of mutation to gain resistance goes to a dangerous level as the enzymes can mutate one at a time. In the case of sulfa-trimethoprim combinations, this is usually not a problem because the combination is administered in a single pill.

MEMBRANE-TARGETING ANTIBIOTICS—POLYMYXIN B AND E (COLISTIN)

Most naturally occurring compounds, synthetic chemicals, or biologically derived molecules with detergent-like properties that can attack bacterial cell membranes will also damage eukaryotic cell membranes, including our own. So, while their broad spectrum of antimicrobial activity makes them highly effective at reducing microbial content from surface areas and preventing infection, particularly in wounds, they are generally viewed as being too toxic for human use inside the body. Consequently, most of these antimicrobials, such as triclosan and hexachlorophene, are usually relegated only for topical applications in antiseptic ointments, creams, toothpastes, or wash solutions, or by prescription in medical devices. We will return to these broad-spectrum membrane-targeting antimicrobials and their continued importance in the battle against antibiotic resistance later in chapter 9. Nevertheless, despite the clear toxic downsides of their use internally, as multidrug-resistant infections have risen in recent years and treatment options have diminished, the use of some of these compounds, such as the polymyxins, for internal applications has made a comeback.

Polymyxins—polymyxin B and E (colistin)

Polymyxins are polycationic peptides produced by Gram-positive *Paenibacillus polymyxa*. Their structures have both charged and lipid-like properties that work by disrupting bacterial membranes. The charged regions interact with lipopolysaccharide in the outer membrane of Gram-negative bacteria, by displacing the calcium (Ca^{2+}) and magnesium (Mg^{2+}) cations from the negatively charged phosphate groups of the phospholipids, which destabilizes the outer membrane. The lipid-like regions of the antibiotics act like detergents to insert into the cytoplasmic membrane and thereby disrupt the integrity of the lipid bilayer, resulting in cell lysis. Polymyxins are not as effective against Gram-positive bacteria, which are protected by their thick peptidoglycan outer layer.

Unfortunately, their detergent-like properties can also damage our cell membranes and cause cytotoxicity. In the 1980s, use of polymyxins as antibiotics was widely discontinued due to adverse damage to kidneys (nephrotoxicity) and nerves (neurotoxicity). Polymyxin B, in combination with other antibiotics such as bacitracin or bacitracin plus neomycin, is often sold as a topical ointment with broad-spectrum antibacterial activity for skin wounds to prevent infections. Polymyxin B is now also used intravenously to treat serious meningitis, pneumonia, sepsis, and urinary tract infections by Gram-negative bacteria for which other antibiotics are not effective. Colistin (polymyxin E) is another polymyxin antibiotic that is currently used as a last-resort treatment for multidrug-resistant Gram-negative infections by *Pseudomonas aeruginosa, Klebsiella pneumoniae, Escherichia coli,* and *Acinetobacter baumannii.*

THE SPECIAL CASE OF TUBERCULOSIS: ANTI-TB DRUGS

Tuberculosis (TB) is an old enemy of the human race, but as discussed earlier in chapters 2 and 3, it is not just a disease of the past. It is currently combining with human immunodeficiency virus (HIV) in Africa and other parts of the world to kill millions of people, and it is once again widespread in the former Soviet Union and in Southeast Asia. In Europe and the United States, TB was once under control, but it has reappeared, and it is being seen more commonly despite the availability of effective therapies. Moreover, in developed countries, a new kind of TB threat has emerged, caused by strains of the TB-causing bacterium *Mycobacterium tuberculosis* that are resistant to one or more of the commonly used anti-TB drugs.

Streptomycin

Streptomycin, as mentioned earlier in this chapter, is an inhibitor of bacterial protein synthesis, and made its debut as one of the first effective TB therapies. Initially, the discovery of streptomycin caused tremendous excitement in the medical community. The excitement soon cooled, however, because streptomycin did not always effect a cure. Remission occurred in some patients. The problem in these patients was that *M. tuberculosis* became resistant to streptomycin during the course of therapy.

In hindsight, this was not surprising, because of the long duration of TB therapy needed to clear the pathogen from the body. It was necessary to administer the antibiotic for many months, in some cases over a year, because once *M. tuberculosis* gets a foothold in the lungs, it is very difficult to eliminate. Within the lungs of an infected person there are two populations of bacteria, one that divides more rapidly and one that divides much more slowly. The acute symptoms of TB are caused by the more rapidly dividing *M. tuberculosis* cells, which are readily killed by the antibiotic. The subset of the population that is growing much more slowly, however, is less susceptible to the antibiotic because lower rates of protein synthesis mean a lower need for ribosomes, which are the target of streptomycin. These bacteria persist and allow the disease to make a comeback even if the infected person's immune system is successful in controlling the initial phase of the disease.

The solution has been to administer a cocktail of three or more antibiotics over a period of at least 6 months, usually more. The antibiotic mixture not only is more effective than a single antibiotic against the rapidly dividing population of bacteria but also is needed to eliminate the slow-growing bacteria. As in the case of sulfa-trimethoprim described above, a mixture of antibiotics can prevent the development of resistance. Despite problems with side effects, the combination therapy has been very effective in curing most patients of TB. In this case, curing means completely eliminating the bacteria from the lungs.

The TB-specific antibiotics

Some of the antibiotics included in TB-specific cocktails have familiar names, such as fluoroquinolones and streptomycin, antibiotics that are used to treat many bacterial infections. Other anti-TB antibiotics have names that are not as familiar because they are TB specific. The three mainstay TB-specific antibiotics are isoniazid, ethambutol, and pyrazinamide. These latter antibiotics are *M. tuberculosis* specific because they inhibit production of a type of bacterial surface membrane lipid, mycolic acid, that is peculiar to the cell wall of *M. tuberculosis* and only a few other types of bacteria (see Fig. 5.3C).

Mycolic acid is responsible for much of the damage to the lungs that occurs when *M. tuberculosis* takes up residence. It contributes to the formation of TB lesions called tubercles that destroy lung tissue and undermine lung function. Thus, inhibition of mycolic acid synthesis seriously interferes with the ability of the bacteria to cause lung damage and at the same time presumably inhibits their growth so that they can be killed by the cells of the host immune system.

Unfortunately, once TB was in abeyance in developed countries (thanks to these drugs), interest waned in the scientific community and funding agencies had less incentive to learn further details of the mechanisms of action of these drugs or how *M. tuberculosis* becomes resistant to them. All of this, of course, changed in the 1990s and 2000s, when TB staged a spectacular comeback in the United States and other developed countries and multiple resistances to the TB-specific drugs began to appear. Heroic and very expensive public health responses to the appearance of these resistant strains in large cities in the United States have brought TB back under control, such that a total of 7,163 cases were reported in 2020; this equates to an average annual decrease of 2 to 3% since 2010. However, this is probably only a temporary victory over a disease that has killed more people than all the wars ever waged and is still out there with the added strength of multidrug resistance. Interestingly, there was a widespread and dramatic 20% decrease in the number of TB cases in the United States between 2019 and 2020. This reduction in TB transmission was presumably due to the mitigation strategies (such as mask-wearing, social distancing, and travel restrictions) that were initially implemented to slow the spread of COVID-19. Sadly, as the pandemic continued and resources globally were hit hard, the previous decline observed in TB incidence worldwide reversed and the rate of new multidrug-resistant TB cases and deaths rose.

Isoniazid. Isoniazid is an antibiotic that specifically inhibits the growth of *M. tuberculosis* and some other closely related species. Activated isoniazid inhibits the synthesis of the mycobacterial lipid mycolic acid, which is a major part of the mycobacterial cell wall. However, it does not inhibit the growth of *Mycobacterium leprae* (the cause of leprosy) or a group of mycobacteria called atypical mycobacteria, which have become an increasing problem for immunocompromised people. One reason for this specificity seems to be, in part, that the administered form of the antibiotic has to be activated by the bacteria before it can carry out its inhibition. Activation occurs because the susceptible species of *Mycobacterium* have an enzyme called catalase-peroxidase that normally protects the bacteria from hydrogen peroxide by converting hydrogen peroxide to water, but the enzyme can also oxidize isoniazid into the active form of the drug. Other strains of mycobacteria do not have this enzyme.

Pyrazinamide. Pyrazinamide has a structure resembling nicotinamide (a form of vitamin B_3), a compound that participates in many bacterial processes, both catabolic and biosynthetic. Like isoniazid, pyrazinamide must first be converted by the bacteria themselves into the active form of the drug. The enzyme that carries out the activation, pyrazinamidase (PZase), is normally involved in nucleotide metabolism, but it can also hydrolyze pyrazinamide to pyrazinoic acid, the active form of the drug. The target of pyrazinoic acid is thought to be an enzyme involved in fatty acid synthesis. Not surprisingly, a mechanism of resistance to pyrazinamide is reduced production of PZase. Thus, an obvious solution would seem to be to use pyrazinoic acid rather than pyrazinamide, but unfortunately, pyrazinoic acid is not taken up by bacteria as well as pyrazinamide, making it less effective.

An interesting feature of pyrazinamide is the way the normal lifestyle of the bacterium makes it susceptible to the drug. *M. tuberculosis* invades the phagocytic cells of the lungs that are supposed to protect the lungs from bacteria. These cells ingest the bacteria, placing them in membrane-coated vacuoles. The interior of the vacuole then acidifies, a step that makes the killing power of the granules that subsequently fuse with the vacuole much greater. *M. tuberculosis* bacteria have developed strategies for avoiding the fate intended for them and can break out of the vacuole and start dividing in the lung cell cytoplasm. Unfortunately for the bacteria but fortunately for us, uptake and accumulation of pyrazinamide into the mycobacteria increases under acidic conditions. Thus, pyrazinamide is thought to act mainly on bacteria that have invaded the phagocytic cells of the lungs. This feature makes it particularly valuable because the bacteria that manage to invade and divide inside the phagocytic cells are the ones that cause most of the lung damage and escape from other parts of the immune system by hiding inside lung cells.

Ethambutol. Until recently, little was known about the mechanism of action of ethambutol. Scientists now believe that ethambutol, like isoniazid, interferes with the synthesis of an important component of the mycobacterial cell walls. This component is not mycolic acid but another mycobacterium-specific cell wall compound called lipoarabinomannan, a complex lipid-polysaccharide molecule (see Fig. 5.3C). Ethambutol is bacteriostatic and works best against actively growing bacteria. Ethambutol has a broader host range than isoniazid and can inhibit the growth of a number of *Mycobacterium* species.

Rifamycins. The most widely used member of the rifamycin family of antibiotics is rifampin (or rifampicin), but its use is restricted to only a few diseases. The most important use of rifampin is in the treatment of TB. Rifampin prevents bacterial growth by inhibiting the enzyme-DNA complex of DNA-dependent RNA polymerase. This enzyme is responsible for using DNA segments of the bacterial genome as a template to produce messenger RNAs (mRNAs), which are subsequently translated into proteins, or noncoding RNA, which have various enzymatic and regulatory functions in the cell.

POINTS TO PONDER

A candidate antibiotic must have certain characteristics for it to be considered an ideal antibiotic for use in treating human infections. Can you name at least four of these traits? What distinguishes a good antibiotic for internal use from one relegated to use only as a topical ointment?

Could a strategy in which two antibiotics that target the same step in peptidoglycan synthesis, such as a combination of a beta-lactam antibiotic and vancomycin, work to counter bacterial resistance? If so, what would be its limitations? Can you guess why pharmaceutical companies have been reluctant to develop such preparations?

Antibiotics that prevent the synthesis of peptidoglycan are usually bactericidal. That is, they kill growing bacteria outright by undermining the integrity of the cell wall. Other antibiotics only inhibit the growth of bacteria, not kill them, by binding to the ribosome and inhibiting protein synthesis. Once the antibiotic is removed from the medium, the antibiotic will become diluted and diffuse away from its ribosome-binding site, thereby allowing the ribosome to resume its function. These antibiotics are called bacteriostatic. Bacteriostatic antibiotics rely on the immune system to clear up the bacteria. Why do bacteriostatic antibiotics work in most people? For what types of patients might bactericidal antibiotics be essential? For what cases might treatment with bactericidal antibiotics not be recommended and bacteriostatic antibiotics be preferred?

6 Bacteria Reveal Their Adaptability and Gain Resistance

Antibiotic use has become so pervasive in modern times and so varied in its application that it is sometimes difficult even for scientists and physicians to keep up with the most recent incarnation of the antibiotic resistance problem. However, the rapidity with which some bacteria have developed antibiotic resistance mechanisms should not have come as a surprise, especially for researchers with any background in the basic biology of bacteria. In this chapter, we will consider some of the ways bacteria counter or subvert the action of antibiotics.

HOW DO BACTERIA BECOME RESISTANT TO ANTIBIOTICS?

In general, bacterial populations become resistant to antibiotics in a two-step process. The first step involves random mutations in individual bacterial cells, either through spontaneous errors during DNA replication or through acquisition of foreign genetic material (DNA). Occasionally, such mutations can cause these individual cells to become more resistant to a particular antibiotic. These mutations can be very subtle, only conferring a slight tolerance for a particular antibiotic compared to the original wild-type bacteria, but they can also be very drastic, especially when acquiring foreign DNA, potentially conferring complete immunity to some antibiotics. In either case, so long as the mutation is not too inherently detrimental to cellular growth, these now-resistant, mutant bacteria will be able to propagate as a subset of the larger bacterial population.

Revenge of the Microbes: How Bacterial Resistance Is Undermining the Antibiotic Miracle, Second Edition.
Brenda A. Wilson and Brian T. Ho.

The second step required for the establishment of antibiotic resistance is for the bacterial population to experience a selective antibiotic pressure—that is, they need to actually be exposed to the antibiotic. Without this selective pressure, mutations conferring resistance to antibiotics do not typically confer any meaningful growth advantage for the bacteria, and therefore the affected bacteria will remain only a minor part of the population. However, once the bacterial population is exposed to the antibiotic, the antibiotic will kill off or inhibit the growth of the bacteria lacking the mutations conferring resistance, while the portion of the population harboring these mutations will continue to survive and multiply, eventually taking over the overall population.

EXAMPLES OF HOW BACTERIA RESIST ANTIBIOTICS

So, what are these mutations that can make bacteria resistant to antibiotics? Mutations conferring antibiotic resistance can be thought of as changes that enable the bacteria to execute one of four main "strategies" to subvert antibiotic treatment. First, bacteria can restrict access of the antibiotic to the interior of the bacterial cell either by physically blocking the agent's entry or by actively exporting it from the cell. Second, bacteria can prevent the action of an antibiotic by enzymatically degrading or chemically modifying it to prevent it from binding its cellular target. Third, bacteria can chemically modify or mutate the cellular target of the antibiotic so that again the antibiotic can no longer bind. Fourth, bacteria can alter their cellular processes or even acquire new ones so that the cellular target disrupted by the antibiotic is no longer essential for bacterial survival. In this section, we will discuss examples of each of these resistance strategies, but it is important to note that although we are considering each resistance mechanism separately here, many multidrug-resistant (MDR) bacteria carry multiple resistance genes and simultaneously employ more than one resistance strategy, which is why they are so difficult to treat.

Preventing access to the target

Changing surface components and membrane properties. The outer membranes of Gram-negative bacteria have pores, formed by proteins called porins, that allow nutrients and small molecules to pass through and gain access to the inner cytoplasmic membrane (see Fig. 5.3B). The openings in different pores have different selectivity in that they only allow certain molecules to pass through, making Gram-negative bacteria inherently more resistant to certain antibiotics than Gram-positive bacteria. Bacterial mutations can alter antibiotic access by either altering how many pores are produced or altering the selectivity of the pores themselves to prevent antibiotic molecules from passing through.

Because penicillin and other beta-lactam antibiotics prevent the last cross-linking step in peptidoglycan synthesis, which occurs in the periplasm or outside on the surface of the cytoplasmic membrane, beta-lactam antibiotics do not need to cross the cytoplasmic membrane to work. Bacitracin also acts on the surface of the cytoplasmic membrane and does not need to enter the cell. However, other antibiotics that must enter the cell's cytoplasm to access their targets can be effectively blocked from acting by preventing transport across the cytoplasmic membrane.

Fosfomycin must enter the cytoplasm of the bacterial cell in order to prevent the synthesis of the sugar-peptide unit that serves as a precursor building block of peptidoglycan. Fosfomycin, like many other antibiotics that act on intracellular targets, must be actively transported across the cytoplasmic membrane by membrane-embedded transporter proteins to access its target. Mutations in genes that encode antibiotic transporter proteins result in antibiotic resistance by restricting access to the intracellular target. For example, *Klebsiella pneumoniae* is a Gram-negative pathogen that is associated with severe hospital-acquired MDR infections. Until recently, fosfomycin was considered one of the limited treatment options for carbapenem-resistant *K. pneumoniae* infections (carbapenem was previously the drug of last resort for these MDR infections). Mutations in the fosfomycin transporter are increasingly seen associated with fosfomycin resistance in carbapenem-resistant strains of *K. pneumoniae.*

Efflux pumps—removing the antibiotic. Because tetracyclines diffuse readily through membranes, bacteria cannot become resistant to tetracyclines by failing to take up the antibiotics through mutation of a transporter. The first mechanism of resistance to be discovered for tetracyclines was a mechanism facilitated by another protein, called an efflux pump, located in the bacterial cytoplasmic membrane, which actively pumps tetracycline out of the bacterial cell (Fig. 6.1).

This bacterial strategy takes advantage of the fact that any compound that binds to a bacterial target within the cell must reach a sufficient threshold concentration for that binding to be effective. For example, an antibiotic like tetracycline that inactivates bacterial ribosomes must pass through the cytoplasmic membrane in sufficient quantities to bind to most if not all of the ribosomes in the cell to halt protein synthesis. A few stray antibiotic molecules are not going to be enough to stop the thousands of ribosomes in each cell from continuing to synthesize proteins. So, if the bacteria can keep the cytosolic concentration of the antibiotic low enough, they will be able to carry on their business as usual. Indeed, most efflux pumps can eject antibiotics from the cytoplasm about as rapidly as the antibiotics can enter the cytoplasm in the first place.

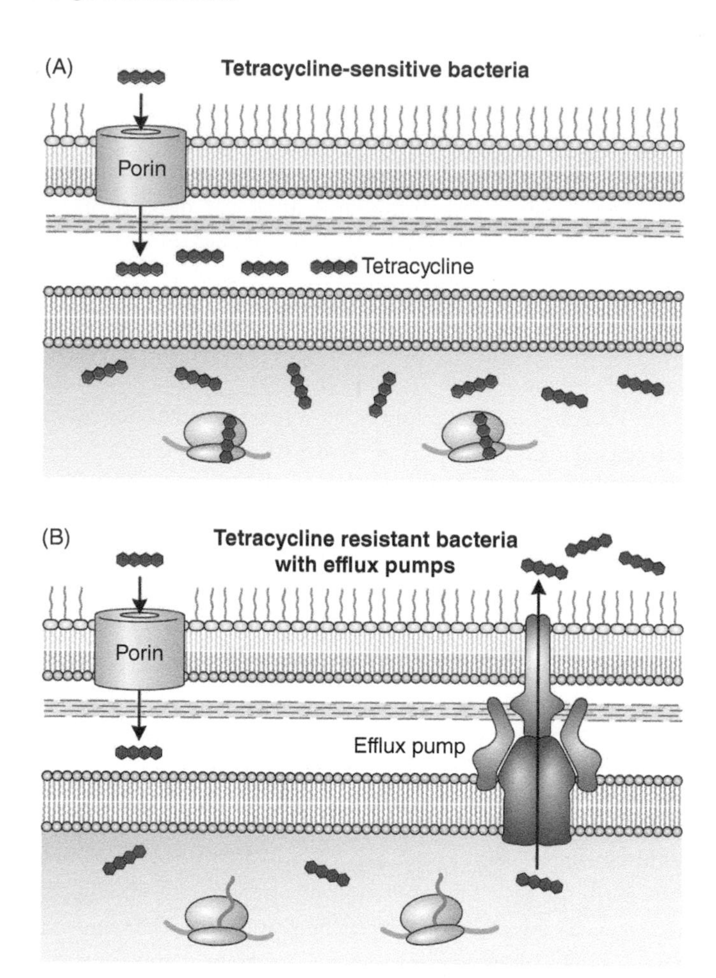

FIGURE 6.1 *Tetracycline resistance mediated through a membrane-embedded efflux pump. (A) Tetracycline can enter cells through porins in the outer membrane and diffusion across the cytoplasmic membrane. (B) Resistant bacteria can produce an efflux pump embedded in their cytoplasmic membrane that can pump tetracycline out of the cell as fast as the antibiotic enters. This keeps the intracellular concentration of the tetracycline too low for it to bind effectively to the ribosomes and inhibit protein synthesis.*

This mechanism of resistance is responsible for resistance to many other antibiotics besides tetracycline, including macrolides, aminoglycosides, streptogramins, and fluoroquinolones. Some efflux pumps are highly specific for a certain type of antibiotic, but others are quite broad in the types of antibiotics that they can remove from cells. These broad-spectrum efflux pumps pose a particularly problematic form of antibiotic resistance that is extremely challenging to combat. Many of the extensively drug-resistant (XDR) clinical isolates of *Pseudomonas aeruginosa* and *Acinetobacter baumannii* ravaging our hospitals have several different kinds of these broad-spectrum efflux pumps, making these pathogens nearly untreatable.

Inactivating the antibiotic before it can act

Bacteria can gain antibiotic resistance by producing enzymes that directly change or destroy the antibiotic before it has a chance to hit its vulnerable targets. Because this type of antibiotic resistance typically occurs through the action of a single enzyme, the genes encoding these proteins are readily transferred among different bacteria, leading to the rapid spread of these antibiotic resistances through bacterial populations.

Beta-lactam antibiotics—beta-lactamases that cleave the beta-lactam ring. The first account of a bacterial strain that was resistant to penicillin was published at about the same time penicillin was introduced into medical use, but it took 3 decades for scientists to finally understand how. In this case, bacteria were protecting themselves through the action of an enzyme called beta-lactamase. This enzyme, as the name suggests, cleaves the beta-lactam ring of penicillin (Fig. 6.2), rendering the antibiotic inactive.

Many beta-lactamases appear to have evolved from the proteins that catalyze the cross-linking of peptidoglycan—the proteins that are the targets of penicillin. When penicillin binds to one of these cross-linking enzymes (aptly named penicillin-binding proteins, or PBPs), the enzyme mistakes penicillin for the two alanine amino acids found at the end of the peptide that will normally participate in the cross-linking reaction to form peptidoglycan. The PBP enzymes can start but not complete the hydrolysis of the beta-lactam ring, resulting in the partially hydrolyzed antibiotic being trapped in the active site of the enzyme and preventing the enzyme from continuing to function, which consequently blocks cell wall biosynthesis. Beta-lactamase has evolved to be able to complete the hydrolysis

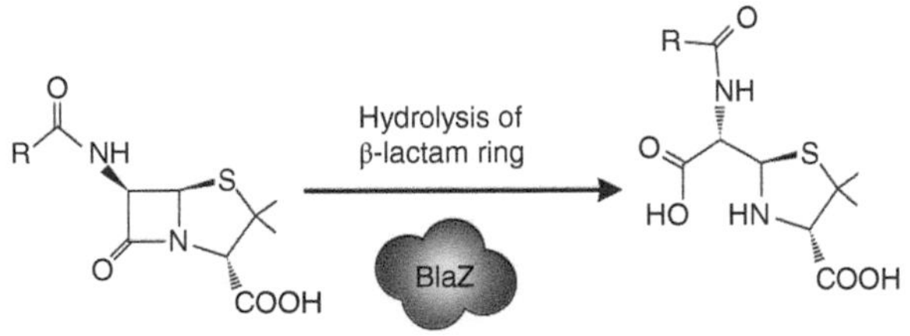

FIGURE 6.2 *Bacterial resistance to beta-lactam antibiotics mediated by beta-lactamases. Bacteria can produce a beta-lactamase enzyme (BlaZ) that destroys the beta-lactam antibiotic by hydrolyzing the beta-lactam ring. The inactivated antibiotic can no longer bind and inhibit its target, the cross-linking enzyme responsible for peptidoglycan biosynthesis.*

reaction and release the penicillin molecule with a broken beta-lactam ring, preventing it from being recognized by the PBPs.

Since the targets of penicillin are located at the cell wall, outside the cytoplasmic membrane, bacteria must secrete their beta-lactamases through the cytoplasmic membrane to have an effect. In Gram-negative bacteria, the beta-lactamase remains trapped in the periplasmic space between the inner cytoplasmic membrane and the outer membrane (see Fig. 5.3B). However, in Gram-positive bacteria, because "outside of the cytoplasmic membrane" is also "outside of the cell" (see Fig. 5.3A), the beta-lactamase can actually diffuse away from the cell, allowing antibiotic resistance to actually be extended to nearby nonresistant bacteria as well.

Chemical modification of antibiotics—enzymes that change the antibiotic's structure. Bacterial resistance can also be mediated by a bacterial protein that chemically modifies the structure of the antibiotic such that it can no longer bind to its target. The main mechanism of resistance to aminoglycoside antibiotics, such as kanamycin and gentamicin, is inactivation by enzymes that change the structure of the antibiotic by adding different functional groups (phosphoryl, adenyl, or acetyl groups) to the hydroxyl or amino groups of the sugar units of the antibiotic (Fig. 6.3A). These modifications block the antibiotic from docking onto the ribosome and thereby prevent the antibiotic from interfering with bacterial protein synthesis.

Likewise, a common mechanism of resistance against chloramphenicol, another antibiotic that targets the ribosome, is acquisition of a gene that encodes an enzyme called chloramphenicol acetyltransferase, which attaches acetyl groups to one or both of the two hydroxyl groups on the antibiotic (Fig. 6.3B). These types of antibiotic-modifying genes are now widely distributed in many different bacteria.

Mutating the target

Although antibiotics have different structures and different targets, one thing they have in common is that resistance to them usually arises as a result of a mutation that disrupts the interaction of the antibiotic with its target. Frequently, that mutation is in the target itself. The balance here for the bacterium is a tricky one. The targets of antibiotics tend to be essential proteins. Thus, mutating them is dangerous for the bacterium, since it could result in an essential protein that is now nonfunctional.

Such mutations are, unfortunately for us, all too easily made by many bacteria. Initially, while the mutation gives the bacterium a survival advantage in the presence of the antibiotic, the mutation may make the bacterium somewhat less able to compete with other bacteria when the antibiotic is absent. However, if the bacterium is given enough time and the antibiotic exposure is sufficiently low to not kill

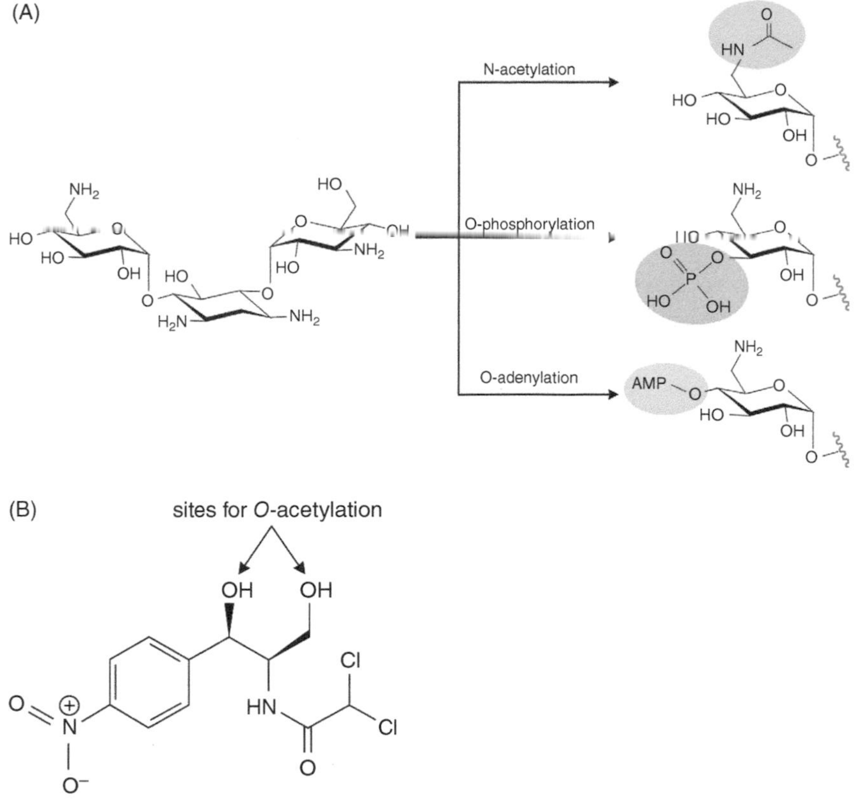

(A)

N-acetylation

O-phosphorylation

O-adenylation

(B)

sites for *O*-acetylation

FIGURE 6.3 *Bacterial resistance through the action of antibiotic-modifying enzymes. (A) Examples of three types of enzyme modifications (acetylation, phosphorylation, and adenylation) that can lead to inactivation of aminoglycoside antibiotics, like kanamycin shown here, result in the antibiotic no longer being able to bind to the ribosome and inhibit protein synthesis. (B) Acetylation of hydroxyl groups on the antibiotic chloramphenicol by the bacterial enzyme chloramphenicol acetyltransferase prevents it from binding to the ribosome.*

the bacteria, additional compensating mutations can occur in other proteins, such that the impact of the original mutation is lessened and the activities of the bacterial cell can be restored to their original optimal levels.

Cell wall targets. Beta-lactam antibiotics bind to and inactivate the PBPs responsible for cross-linking peptidoglycan during cell wall biosynthesis. One mechanism for bacteria to become resistant to beta-lactam antibiotics is by carrying genes for mutant PBPs that do not bind beta-lactam antibiotics. This type of resistance is most commonly associated with Gram-positive pathogens such as *Streptococcus pneumoniae* and *Staphylococcus aureus,* but it has been observed in some Gram-negative pathogens as well.

Protein synthesis targets. Some bacteria prevent antibiotics from harming their ribosomes by modifying one or more components of the ribosome to prevent the drug from binding. As mentioned above, this strategy can be somewhat dangerous for bacteria, as mutating or modifying essential components or processes can be detrimental to the bacterium. Yet bacteria still succeed in finding mutations that allow them to resist the action of the antibiotic while still carrying out their normal life processes. For example, one way that bacteria become resistant to aminoglycosides like streptomycin is by mutating the ribosomal protein to which they bind. These mutations cause the antibiotic either to no longer bind or to bind in such a way that the ribosome is no longer inhibited from undergoing protein synthesis. Fortunately for us, antibiotic resistance through ribosomal mutations is not seen often in clinical settings because mutating ribosomal proteins typically makes the bacteria less fit in the absence of the selective pressure created by the antibiotic.

DNA and RNA targets. One type of erythromycin resistance gene, called *erm* genes, is also called MLS-type resistance because these genes encode enzymes (RNA methylases) that modify the ribosomal RNA (rRNA) target of macrolides (M), lincosamides (L), and streptogramins (S). Bacterial resistance to the MLS group of antibiotics illustrates how bacteria, in a single stroke, can evolve a mechanism of resistance, in this case methylation of the rRNA, that confers resistance to more than one class of antibiotic simply because all of these antibiotics bind to the same rRNA molecule. This type of resistance has been observed mostly in Gram-positive bacteria and in *Bacteroides* species, which are members of the normal gut microbiota.

Bacteria can become resistant to fluoroquinolones by making one or a few mutations in one of the genes that encode the DNA gyrase protein, so that the antibiotic no longer binds to the mutant enzyme. Similar resistance to rifampin results from mutations in the RNA polymerase that prevent rifampin binding. Because these mutations arise from only a single amino acid change, the frequency with which they occur is as high as 1 in 10 billion bacteria. While this mutation rate may intuitively seem low, it is important to remember that bacteria are often present in the human body in quantities that are hundreds to thousands of times higher than that, and once the mutant appears even a single time, with sufficient selective pressure, that mutant will be able to supplant the whole population.

Bypassing the antibiotic's action

While it is not common, some bacteria gain resistance by preventing activation of the antibiotic or by bypassing the need for the antibiotic's target. We include here a few important examples of antibiotics for which these mechanisms are used by bacteria to resist them.

Failure to activate the antibiotic. Metronidazole and isoniazid were two antibiotics mentioned in chapter 5 as having the unique feature that they must first be activated by the bacterium before they can kill the bacterium. Metronidazole must first be reduced before it can attack DNA. Mutations that decrease the expression of the flavodoxin and ferredoxin enzymes that are required to convert metronidazole into its active form lead to resistance. Similarly, mycobacterial mutations in the catalase-peroxidase gene that inactivate the enzyme, reduce its expression, or otherwise prevent it from being made result in mycobaterial resistance to isoniazid, which in turn prevents the antibiotic from inhibiting biosynthesis of the cell wall component mycolic acid.

Acquiring new processes that avoid the antibiotic's target. Resistance to trimethoprim and sulfonamides usually occurs from spontaneous mutation in the enzymes involved in folate biosynthesis. Some resistant strains of *S. aureus* emerge after long-term treatment with the trimethoprim-sulfamethoxazole combination drug. These mutant strains bypass the action of the sulfa-trimethoprim drug by no longer requiring folate (needed for thymidine biosynthesis), but instead, these bacteria are now dependent on direct uptake of thymidine (DNA precursor) from the environment.

THE RISE OF RESISTANCE TO CURRENT ANTI-TB DRUGS: CONNECTING THE DOTS

Tuberculosis (TB) has always been a major cause of infectious disease deaths worldwide but had been thought to be under control in developed countries. For decades, the different cocktails of anti-TB drugs remained highly effective. While some strains of *Mycobacterium tuberculosis* did become resistant to individual drugs, this phenomenon was rare enough that the combination of multiple drugs was still effective. However, starting in the 1980s, multidrug resistance began to appear more and more often. By the 1990s, drug-resistant TB had become a serious problem in the United States and other developed countries. To have TB not only stage a comeback in developed countries but also emerge as increasingly resistant to previously effective treatments was troubling for many health officials.

What happened? Once again, complacency had raised its dangerous head. In the past, when TB was taken seriously in the United States, a public health infrastructure existed that sought and treated the disease very aggressively. Most people do not comply very well with a drug regimen that forces them to take several pills daily or two to three times a week for more than 6 months. Also, the drugs have side effects such as nausea for many people or hair loss in others, a feature that makes compliance worse if patients take the drug unsupervised. The solution was to compel TB patients to come regularly to a clinic or to have health workers visit them in their homes. The pills were given to the patient, and the health worker watched to make sure that all of them were taken. This strategy has been called DOTS (for directly observed therapy short course). DOTS was highly effective, and as long as this program was in place, resistance did not arise and virtually all treated people were cured of the disease.

Starting in the 1970s, however, public officials in the United States decided that since TB was under control, the relatively small number of cases did not merit the expensive infrastructure necessary to sustain a DOTS program. Dismantling of the anti-TB infrastructure began. Health workers were no longer recruited for DOTS programs, X-ray machines used to diagnose TB were given away by community health institutions, and doctors were not trained as aggressively in how to diagnose and treat TB. Even worse, work on developing new anti-TB drugs virtually ceased, as did efforts to understand how the existing anti-TB drugs worked.

All might have been well, at least for a while, if it had not been for some seemingly unrelated developments, such as the rise in the prison population, the increase in the number of homeless people housed in crowded shelters, and the advent of HIV. In crowded conditions, TB spreads rapidly, so the crowded jails and homeless shelters were a TB disaster waiting to happen. Moreover, HIV-infected people were less able than uninfected people to fight off microbial infections and had to be treated much longer for TB (a year or more rather than just 6 months). Any HIV-positive patient who progressed to AIDS was particularly vulnerable to TB since the immune cells needed to contain TB were destroyed by the active virus. Abuse of alcohol and other drugs was another factor that decreased even further the likelihood that a TB patient would comply faithfully with the long and complex therapeutic regimen.

Unsupervised infected people might take one drug for a while and then another, but usually only irregularly. Recall that the reason the anti-TB drugs are given as a cocktail is that experience had shown that use of just a single drug often led to the development of resistance. *M. tuberculosis* became resistant first to one and then another of the anti-TB medications. A frightening example of where this was heading can be seen in the appearance in the 1990s of a strain of *M. tuberculosis* called the W strain. This strain swept through first New York City and then other large cities. The W strain not only was resistant to most of the anti-TB antibiotics but was as virulent as other *M. tuberculosis* strains, if not more so.

The good news is that research on the development of new anti-TB drugs is under way and U.S. health care officials have reinstated DOTS. Draconian intervention measures have brought outbreaks of drug-resistant strains of *M. tuberculosis* under control. Most healthy people who are infected with *M. tuberculosis* do not immediately develop the full symptomatic form of the disease. Rather, they harbor the bacteria in their lungs for years. In some cases, stress on a person's immune system from age, cancer chemotherapy, or any of a variety of other sources allows the bacteria to break out of their dormancy. The result is called "reactivated TB," which is just as deadly and infectious as primary TB. Many of the people who were infected with the W strain and developed symptomatic disease died. Yet there are many more who carry that strain in their lungs.

IS IT ALWAYS THE BACTERIUM'S FAULT? OTHER REASONS FOR TREATMENT FAILURE

Although the emergence of bacteria that have developed new ways to resist antibiotics is unquestionably a serious clinical problem, the failure of an antibiotic to cure your condition may not always be due to bacterial resistance. It is possible that moving to the newest high-powered (read "expensive") antibiotic may not always be the best solution to a treatment problem. How can an antibiotic treatment fail for reasons other than the presence of bacteria that resist the action of the antibiotic? A major reason for apparent antibiotic failure is misdiagnosis of the infection or application of the antibiotic as a broad-spectrum cure-all without first ascertaining that the disease is caused by a bacterium.

If the patient's condition is caused by a virus or a fungus, for example, and not by a bacterium, antibiotics will have no effect. An example of this can be seen in the case of some urinary tract infections. The vast majority of urinary tract infections are caused by bacteria. Thus, antibiotics are a standard treatment for the painful urination and fever associated with such an infection. In some people, however, urinary tract infections are caused by the yeast *Candida albicans*. Such an infection cannot be treated successfully with an antibiotic. Similarly, misdiagnosis of a genital tract infection, which usually is caused by bacteria but in some cases may be caused by fungi or protozoa, can lead to treatment failure if an antibiotic is prescribed as the treatment.

It is important to note, however, that a patient can go from feeling minor discomfort to being dead in just a couple hours if certain bacterial infections go untreated and bacteria are able to enter the bloodstream. As such, doctors do not always have the luxury of properly diagnosing what bacterium is infecting a patient or even whether it is actually bacteria that are causing the problem. It is sometimes safer to just prophylactically treat with antibiotics to prevent the worst-case scenario, even if a large percentage of the time the antibiotics were unnecessary. This challenge is one of the major motivations driving development of new rapid diagnostic techniques to help make sure that antibiotics are only used when they will be effective.

Another potential problem might arise through use of an antibiotic that has the wrong pharmacokinetic properties—a fancy way of saying that the drug does not get where it is needed. For example, an antibiotic that does not penetrate the blood-brain barrier (the membranes that cover the brain and spinal cord and separate spinal fluid from blood) will not do much for a meningitis patient, who has bacteria in the cerebrospinal fluid. Generally, this is not a common problem, because most physicians are well trained in the pharmacokinetics of the drugs they use. However, sometimes it is out of their hands due to incomplete information in the medical community. For example, for a long time, little was known about what types of antibiotics were most useful in treating bacteria growing in abscesses, where dead tissue inhibits the penetration of some antibiotics. Now there is a better understanding of what antibiotics work best or whether removing the dead tissue is going to be essential before any antibiotic can enter the site.

Patients can also play a role in treatment failure. Failure to take the full course of an antibiotic can leave viable bacteria to make a comeback when the patient stops taking the drug. Many people have difficulty complying with certain drug regimens, especially when they experience adverse side effects such as nausea. Not only does intermittent taking of pills, or cessation of therapy entirely, leave live bacteria in the patient's body, it also increases the likelihood that the remaining bacteria may become resistant, rendering further attempts at therapy ineffective.

THE ORIGINS OF BACTERIAL RESISTANCE GENES: A PUZZLING BUT RELEVANT MYSTERY

How do bacteria develop the special genes that make them resistant to antibiotics? Scientists are trying to answer this question by determining the DNA sequences of the genes that confer resistance and asking whether they resemble any other genes that were present before the resistance problem emerged. If there are similarities, perhaps scientists can construct a "fossil record" of the evolution of resistance genes that reveals how they arose.

Also, did antibiotic-resistant bacteria emerge only after antibiotics were introduced into widespread clinical use or were they present before antibiotics were widely used? There is now sufficient evidence to support that at least some antibiotic resistance genes were present long before humans began to produce and use antibiotics. This was an important finding because it forced us to rethink antibiotic discovery strategies. Since many of our antibiotics are derived from natural products, emergence of resistance may be less about evolution of new resistance genes and more to do with the spread of existing resistance genes already present in nature.

One prevailing explanation for the existence of bacterial resistance genes before antibiotic use became widespread is that antibiotics were the first forms of germ warfare. That is, bacteria and fungi that produce antibiotics likely have long been using these molecules to clear competing bacteria from their vicinity, granting them unmolested access to the resources found in that particular location. Antibiotic resistance genes may have simply evolved as a counter to these naturally-produced antibiotics. A problem with this undeniably attractive idea is that we do not typically detect a lot of antibiotics in soils where microbes that produce antibiotics in the laboratory are normally found, which is consistent with the fact that it is usually necessary to subject a possible antibiotic-producing microbe to mutagenesis to increase production of an antibiotic to usable levels. That said, recent studies have found that part of the reason antibiotics are hard to detect in soil is that they tend to bind tightly to soil particles or form complexes with other materials. Additionally, these soil microbes harbor a number of different antibiotic resistance genes, strongly suggesting that they do encounter antibiotic pressure.

Why is this debate of more than academic interest? Scientists are desperately designing strategies for controlling and slowing the increase in bacterial resistance to antibiotics on the assumption that the only selection for resistant bacteria is the use of antibiotics in modern medicine or in modern agriculture. What if there are selective pressures other than antibiotic use by humans that are driving the evolution of antibiotic-resistant strains of bacteria? Ancient selective pressures might be enhanced by increased human influences on the environment such as pollution or the release of antibiotics into nature. There is even some evidence that heavy metals and other pollutants may be selecting for bacteria resistant to antibiotics. Such selective pressures might not be major contributors to the current rise in incidence of resistant strains, but they could be generating a constant source of small numbers of resistant bacteria, whose numbers increase dramatically if an antibiotic selection is applied. An understanding of such non-antibiotic selective pressures, if they exist, has great practical importance in the battle to save antibiotics.

POINTS TO PONDER

This chapter lays out the usual and most prevalent bacterial resistance mechanisms, the ones for which there is abundant evidence: inactivating the antibiotic, reducing the antibiotic concentration by pumping it out of the cell, and modifying the target of the antibiotic. Given that most of the antibiotic targets are enzymes or other proteins that perform essential bacterial functions and given that many antibiotics act on targets in the cell cytoplasm, can you think of other ways bacteria might become resistant to antibiotics? Just because other mechanisms have not been described yet does not mean they do not exist. There is still a lot we do not know about the seemingly endless ability of bacteria to evolve around antibiotics.

The avoparcin debacle mentioned in chapter 3 occurred in Europe at the same time as the intense controversy over the safety of genetically modified crops. Europeans made it clear that they wanted to be super-safe when it came to the food supply. Yet they clearly applied a different standard to antibiotic resistance genes that had ended up in the plant cells as a consequence of cloning than to the agricultural use of avoparcin. How could such a situation develop? Why would the same public react in such different ways to the problem of antibiotic-resistant bacteria?

Now that we know that some antibiotics (e.g., macrolides, lincosamides, and streptogramins) can cross-select for resistance to more than one class of antibiotics, should the criteria for approval of antibiotics for use in agriculture be changed, or should approval continue to be decided one antibiotic at a time? The answer to this question is not as obvious as it might seem, because cross-resistance usually does not apply to all members of all the antibiotic families involved. That is, erythromycin-resistant bacteria that have acquired an *erm* gene (MLS-type resistance) do not become resistant to all macrolides, lincosamides, and streptogramins. Similarly, the broad-spectrum efflux pumps that pump out more than one class of antibiotic are generally not universally effective. Yet.

7 The Spread of Antibiotic Resistance

As we discussed in the previous chapter, the most common form of antibiotic resistance originates from mutations in a bacterium that either reduce its susceptibility to an antibiotic or enable it to destroy the antibiotic. However, such mutations do not come cheaply to the bacterium, as every mutation carries a risk that it could kill the bacterium or render it severely crippled. How then is antibiotic resistance such a widespread and prevalent issue? Simply put, bacteria are more than happy to share with each other. As soon as an individual bacterial cell develops antibiotic resistance genes after rolling the genetic dice enough times, the fruits of that labor can easily be shared with not only its siblings and progeny but also other nearby bacteria. Unfortunately for us, this "generosity" has greatly accelerated the global antibiotic resistance crisis we now find ourselves in. In this chapter, we will consider some of the ways that bacteria share their resistance genes with each other and some of the ways we humans facilitate this spreading process.

HOW BACTERIA RAPIDLY ACQUIRE AND SPREAD ANTIBIOTIC RESISTANCE

Even though bacteria that have mutated to gain resistance can then mutate further to become more comfortable with their newly resistant state, as discussed in chapter 6, a much easier way for bacteria to become quickly resistant to an antibiotic is to acquire an already-existing antibiotic resistance gene from another bacterium. However, unlike higher-order plants and animals, bacteria primarily reproduce

Revenge of the Microbes: How Bacterial Resistance Is Undermining the Antibiotic Miracle, Second Edition.
Brenda A. Wilson and Brian T. Ho.

clonally through binary fission, with each newly formed daughter cell being an exact genetic duplicate of the parent. As such, genetic exchange is generally not an inherent part of their replicative life cycle. Instead, bacteria can engage in a form of bacterial sex called conjugation, during which genetic material can be transferred from one bacterial cell to another. Furthermore, bacteria are quite promiscuous, with these genetic transfer events often crossing species and genus lines. Consequently, conjugation and other similar gene sharing processes enable antibiotic resistance genes to spread rapidly among bacterial populations.

Gene sharing—resistance acquisition via horizontal gene transfer

There are three routes through which bacteria can acquire new genetic material. Collectively, microbiologists call this type of DNA acquisition horizontal gene transfer (HGT) because it occurs not vertically through reproduction (i.e., not from parent to offspring) but rather through gene sharing between two cells.

The first, transformation, is the process by which some bacterial cells take up extracellular DNA segments from the environment through special membrane complexes. Bacteria with these DNA uptake systems are said to be "competent for transformation" or just simply "competent." When other bacterial cells lyse and release their DNA into the environment, competent bacteria can take up the released DNA and incorporate it into their genomes. The second process is called transduction and is mediated by bacterial viruses called bacteriophages or just phages. Analogous to eukaryotic viruses, a phage will invade a bacterial cell and use its replication and protein production machinery to reproduce. During this process, on rare occasion, some phage particles will be accidentally packaged with a random portion of the host genome instead of the phage genome. When these mispackaged particles go on to infect other bacteria, rather than infecting the cells like normal, they instead deliver a chunk of the previous host's genome. The third route is conjugation, which as described above is a form of bacterial sex where genetic material is directly transferred from one cell to another.

All three types of HGT can mediate the transfer of antibiotic resistance genes, but transformation and transduction both have inherent limits to their scope. Most bacteria tightly regulate their competence behaviors and extracellular DNA can be readily degraded, so transformation occurs only under special conditions. Meanwhile, most phages have a limited host range, meaning that they are typically very species specific. On top of that, the mispackaging events needed for transduction are extremely rare. Conjugation, on the other hand, only requires that two bacteria come into direct contact with each other. Since DNA is directly transferred between the cells, the extracellular stability of DNA is a nonissue. Additionally, many conjugation systems have an extremely broad host range, meaning that they

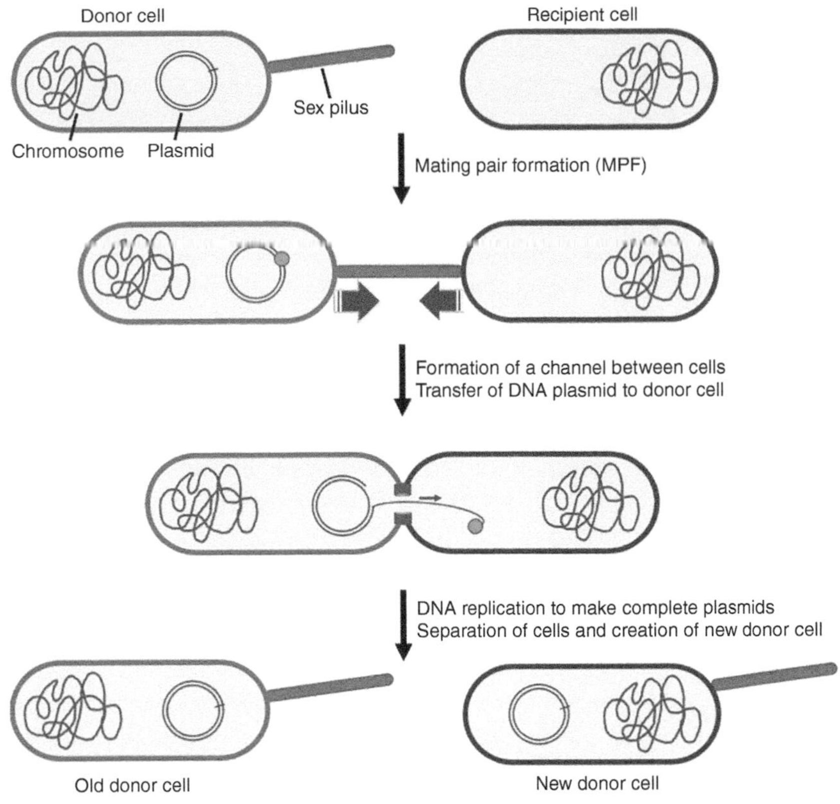

FIGURE 7.1 *Mechanism of horizontal gene transfer between bacteria through conjugation of DNA plasmids. Shown are the steps involved in transfer of a DNA plasmid via bacterial conjugation from a donor cell to a recipient cell and conversion into a new donor cell.*

can transfer DNA into a wide range of cells—some can even deliver DNA into fungi, plant, and animal cells.

Bacteria capable of initiating conjugation, called donor cells, employ a specialized, multiprotein nanomachine to mediate the DNA transfer process (Fig. 7.1). First, the donor bacterium extends a long, tube-shaped appendage called a mating or sex pilus out of the cell. When this pilus contacts a recipient cell, it adheres to its surface. The pilus is then retracted, pulling the recipient in close to the donor in what is called a "mating pair." A channel is then formed between the two cells, and DNA is transferred from the recipient to the donor. Once transfer is complete, the channel between the mating pair is resolved and the two cells are free to go their separate ways. Another feature worth noting about most conjugative systems is that the protein machinery that mediates the DNA transfer is typically encoded by genes located on the DNA plasmid that is transferred. As a result, recipient cells typically become conjugative donors themselves, helping to further spread the conjugated DNA plasmid to the next recipient.

Most conjugated DNA exists in the form of plasmids. These segments of DNA can be either circular or linear, but most importantly, they are able to autonomously replicate, meaning that they can replicate inside a cell independent of the bacterial chromosome. Usually, the cell's machinery (e.g., DNA polymerase and helicase) is used to do this, but the extremely broad-host-range plasmids often encode their own replication initiator proteins. As mentioned above, the conjugation machinery is usually encoded by genes on transferable DNA. However, that does not mean that all transferable DNA encodes its own conjugation machinery. In fact, any DNA that carries a special mobilization sequence can be transferred by the conjugation machinery. Scientists regularly make use of these systems in the laboratory to genetically modify research organisms.

Plasmids have the capacity to carry a wide array of different antibiotic resistance genes. Many commensal bacteria that reside within the gut or other human microbiomes can harbor these plasmids. As such, they can serve as potential "reservoirs" of antibiotic resistance genes that passing pathogens can tap into.

Transposons

In addition to transferring between cells through HGT, many antibiotic resistance genes have the capability of moving around within an organism as well. Transposons, sometimes called "jumping genes," are genetic elements that can relocate or transpose themselves from one piece of DNA to another. They can jump between sites on bacterial chromosomes or between sites on plasmids and the chromosome. When they insert into conjugative plasmids, they can then also move throughout bacterial populations along with the plasmid. The transposon DNA usually encodes the enzymes that mediate the jumping process, but also frequently encodes additional genes like antibiotic resistance genes.

The mobile nature of transposons means that they can help genes overcome some of the limitations of HGT. For example, during transformation, for acquired DNA to be retained, the extracellular DNA that is taken up must be integrated into the bacterial chromosome, which requires some degree of sequence similarity between the acquired DNA and the existing genome. If the acquired DNA cannot be stably integrated into the bacterial chromosome, it will eventually be lost. Transposons enable antibiotic resistance genes to completely bypass this integration requirement by simply hopping from the acquired DNA straight into the chromosome. Similarly, conjugative plasmids or phages that can enter cells but for whatever reason cannot replicate in their new host would normally lose this newly acquired DNA. But if the plasmid or phage carried a transposon, the transposon could still hop directly into the bacterial chromosome.

Selecting for and maintaining multidrug resistance

Given our discussion so far, it should be clear that antibiotic resistance genes are often carried on mobile genetic elements that can spread through bacterial populations. But if these genes are mobile, you might be wondering why antibiotic resistance genes cannot be lost just as easily. Easy come, easy go, right? Unfortunately, it is not so simple. Because antibiotics are so good at eliminating bacteria lacking resistance, their presence creates an extremely strong evolutionary pressure to force resistance to be retained within bacterial populations. So why not remove that selective pressure? Surely, we can just rotate which antibiotics we use to prevent bacteria from being exposed to any single antibiotic long enough to develop resistance, right? Well, again, it is not so simple. As it turns out, there are several different factors that drive antibiotic resistance gene maintenance, but perhaps the most important is that many of the most common antibiotic resistance mechanisms confer resistance to multiple different drugs. Therefore, if bacteria encounter any of one these drugs, resistance to all of them is retained.

Several of the drug resistance mechanisms discussed in chapter 6 fall into this category of multidrug resistance. First, enzymes conferring resistance by degrading or chemically modifying an antibiotic often can target all antibiotics belonging to the same drug family. For example, many of the enzymes conferring resistance to ampicillin also confer resistance to other beta-lactam drugs like penicillin or amoxicillin.

Other resistance mechanisms can also confer resistance to antibiotics of different classes, where there is no structural similarity. The simplest of these are the broad-spectrum efflux pumps, which can remove several different unrelated small molecules from the bacterial cell. If selective pressure is applied for the bacterium to retain the pump, the bacterium will retain resistance to all drugs removed by that pump.

Lastly, resistance mechanisms that involve altering the antibiotic drug target can often simultaneously affect multiple different drugs that all bind to the same site on the target. In chapter 3, we mentioned that the macrolide tylosin has been one of the antibiotics used on farms across the United States that has contributed to the antibiotic resistance problem for human disease. The macrolide tylosin has been widely used in swine production to prevent bacterial respiratory diseases. The problem is that, although tylosin has a different name and somewhat different structure from erythromycin, it is similar enough to erythromycin to select for resistance to erythromycin and other macrolides used for treating humans. Tylosin also selects for resistance to lincosamides and streptogramins. The reason is that the macrolides, lincosamides, and streptogramins (the MLS group mentioned in chapter 6), despite their different structures, all bind to the same or overlapping regions of the ribosomal RNA (rRNA) molecule that forms part of the large subunit of the bacterial ribosome. Thus, a mutation in this rRNA molecule or acquisition of a bacterial enzyme that can methylate a key base on this rRNA molecule simultaneously confers resistance to all three classes of antibiotics by reducing their binding to the ribosome.

To make matters worse, in addition to cases where a single resistance mechanism is broadly acting, when multiple resistance genes are added stepwise onto the same mobile DNA element, they become genetically linked. When this happens, not only are they transferred together between bacteria simultaneously in a single event, but they can effectively help select for the retention of all the linked genes. To put it more concretely, what this means is that if a plasmid contains both a beta-lactamase gene and a kanamycin-modifying acetyltransferase gene, exposure of a bacterium harboring this plasmid to penicillin will select for not only maintenance of the beta-lactamase and resistance to other members of the beta-lactam antibiotic class but also resistance against kanamycin and other aminoglycoside antibiotics.

ANTHROPOGENIC FACTORS THAT CONTRIBUTE TO THE SPREAD OF ANTIBIOTIC RESISTANCE

An inevitable consequence of the use of antibiotics, especially at subtherapeutic levels that do not eliminate the bacterium, is the selection for antibiotic resistance genes and the survival and dissemination of the resulting resistant bacteria. A corollary that follows this well-established connection is that we all, however inadvertently or unknowingly, have contributed to the rise and spread of the antibiotic resistance problem through our actions. Here, we will consider a few examples of ways in which this has occurred.

Use of antibiotics in human medicine and resistance gene transfer

As indicated by an example given in chapter 5, orally administered tetracycline is still widely used by dermatologists to treat acne and another skin condition, rosacea (reddening of the skin that looks like a permanent blush). People taking tetracycline for a skin condition often take it for long periods of time, sometimes for years. While there is evidence that all types of tetracycline resistance (efflux pumps, ribosome target mutation, and enzymatic drug inactivation) have evolutionary origins from environmental sources, long-term antibiotic use has likely contributed to the increase in tetracycline-resistant bacteria found among pathogens circulating in hospital settings, as well as the normal resident bacteria in our bodies. It is highly likely that the long history of tetracycline usage for medical applications and its widespread deployment for agricultural applications have driven the increased prevalence of tetracycline resistance.

When a physician uses an antibiotic to treat a person for a specific infection, they also "treat" the patient's normal resident microbiota, the bacterial populations that normally occupy the patient's skin, mouth, lung, and intestinal and urogenital tracts. Since the bacteria in these "off-site" locations are not always exposed to the full inhibitory concentration of the drug, antibiotic resistance can readily develop in these populations. Usually, these bacteria are neutral or even protective, but under certain circumstances, such as during surgery, opportunistic members of this normally innocuous bacterial population can cause serious infections if they enter areas of the body from which they are normally excluded. Typically, these opportunistic pathogens are held at bay by antibiotics, but if they become passively resistant to antibiotics due to previous antibiotic treatment, suddenly a routine surgical operation becomes a life-threatening endeavor.

Unfortunately, it is not just resident microbiota that are of concern. Once resistance has been acquired within bacterial populations, it is only a matter of time before that resistance ends up in a pathogen. By the early 2000s, the spread of antibiotic resistance had left the drug vancomycin as the last line of defense against many clinical isolates of *Staphylococcus* and *Streptococcus* species, both of which can cause overwhelming fatal systemic infections. Since vancomycin began to be used heavily in human medicine years earlier, most strains of these multidrug-resistant (MDR) bacteria remained reassuringly susceptible to the drug, but this barrier finally started to crack when some *Enterococcus* species were reported to be vancomycin resistant.

Enterococcus species are Gram-positive bacteria related to staphylococci and streptococci found in the colons of nearly everyone. Normally, enterococci do not cause infection, but if they escape from the colon, they can cause serious wound and bloodstream infections. Some hospitals reported very bad experiences with postsurgical *Enterococcus* infections in their intensive care wards. Even in this case, however, it was somewhat reassuring, in a backhanded way, that *Enterococcus faecium*, which seemed to be the species least able to cause infection, was the one that was most likely to become resistant to vancomycin.

Although it had been shown before in the laboratory that *Staphylococcus aureus* could become resistant to vancomycin by acquiring vancomycin resistance genes via conjugation, when transfer of vancomycin resistance genes to *S. aureus* did not happen immediately in hospitals, physicians began to hope that *S. aureus* would remain obediently susceptible despite repeated encounters with the resistant enterococci in the gut. In 2003, this hope was shattered when the first clinical case of a vancomycin-resistant strain of *S. aureus* was reported. The strain was isolated from a patient receiving kidney dialysis. A mixture of both *S. aureus* and *E. faecium* was isolated from the patient's bloodstream. In this case, some of the *S. aureus* isolates were susceptible to vancomycin, while others were resistant. Further investigation provided evidence that the *E. faecium* isolates from the patient were all vancomycin resistant and likely had transferred the resistance gene to *S. aureus*, possibly within that same patient.

This was an ominous development. Vancomycin-resistant *S. aureus* strains with complete resistance to vancomycin have become more prevalent in recent years and have spread to many different cities and countries around the world. There is a tiny silver lining in this very dark cloud. So far, while there are MDR strains of *S. aureus* and other pathogens that are resistant to nearly all antibiotics, the so-called extensively drug resistant (XDR) strains, these are still rare. The caution here is that even if there are still one or more antibiotics that will work, much more laboratory diagnostic work must be done to determine which antibiotic to use for such a patient. Unfortunately, for patients with sepsis (systemic bacterial infection), rapid treatment is paramount. The longer the delay before the bacteria can be brought under control, the greater the risk that the patient may sustain irreversible damage to important organs such as the lungs, kidney, and brain from the progression of septic shock. Some patients may even die if the shock process has gone too far for an antibiotic to be effective, even if an antibiotic that is effective in the laboratory can be identified.

Hospital-acquired MDR infections and the ESKAPE pathogens

Nowhere is the threat of MDR infections grimmer than in hospital settings, where antibiotic treatment is essential to prevent dire clinical outcomes among the highly vulnerable patient population. Unfortunately, the frequency of hospital-acquired infections by MDR pathogens is escalating at alarming rates. The resistance profiles of these pathogens may vary somewhat, depending on the specific pathogen, patient susceptibility, hospital setting, and specific antibiotics administered, but the MDR arises generally through mechanisms that we have already mentioned, including subtherapeutic dosing that selects for resistance; use of different kinds of antibiotics in the clinical setting; spread through HGT among circulating strains; and close contact with many vulnerable patients in hospital or health care settings, such as nursing homes, daycares, or outpatient clinics. Another contributing factor leading to the rise in MDR strains in clinical settings is the lack of inexpensive, rapid, and comprehensive diagnostic tests to identify not only the bacterial pathogens but also their resistance genes. The urgent need under such circumstances to immediately implement a treatment plan often necessitates the use of broad-spectrum antibiotics, which in turn leads to simultaneous selection of multiple resistances, as well as additional health complications due to loss of normal resident microbiota.

A group of six Gram-positive and Gram-negative MDR pathogens have become notorious not only for causing deadly hospital-acquired (nosocomial) infections but also for increasing MDR to most of the classes of antibiotics commonly used to treat infections by these pathogens. These MDR pathogens have been dubbed "ESKAPE" pathogens for the first letter of the name of each pathogen: *Enterococcus faecium, Staphylococcus aureus, Klebsiella pneumoniae, Acinetobacter baumannii, Pseudomonas aeruginosa,* and *Enterobacter* species. These pathogens are responsible for most of the more serious nosocomial infections seen in recent years. Particularly problematic is the appearance of more and more of these clinical isolates displaying extensive drug resistance (XDR), rendering such infections essentially untreatable. ESKAPE pathogens, which can evade or "escape" normal antibiotic treatment regimens, are placing an enormous financial and resource burden on global health care systems, and consequently will be a tremendous future challenge to overcome.

Agricultural use of antibiotics and environmental sources of resistance

As described in chapter 3, tons of antibiotics are used in animal husbandry every year. A substantial amount of those antibiotics (no one can agree exactly how much) is used to enhance the growth of food animals, especially pigs and chickens. After the negative experience with avoparcin (discussed further in the next section), farmers in Europe have largely given up the use of antibiotics to promote growth and have restricted them to prophylactic use. There is a key difference, however, between the way animal husbandry is practiced in the United States and Europe. In the United States, consumers have demanded cheap meat prices. They also want to have animal production facilities confined to the smallest space possible, leaving more land available for houses and parks. The result has been an industry that runs farms on which animals live in crowded conditions, conditions that favor the spread of infectious diseases. By comparison, European farms house far fewer animals at lower densities.

What does this little excursion into animal husbandry have to do with the spread of antibiotic resistance? The answer lies in the series of steps that might link the use of large quantities of antibiotics in agriculture with increasing resistance to antibiotics by bacteria that cause human disease. Antibiotic-resistant bacteria are selected in the intestinal tracts of animals that are fed antibiotics. These resistant bacteria contaminate the carcass during slaughter and enter the food supply. There is now ample evidence that this contamination occurs not only in meat products but also water sources and other food crops. Undercooking enables survival of the bacteria, which can then enter the consumer's colon. In the colon, the incoming bacteria encounter the bacteria that normally reside in the human colon. Since incoming bacteria are adapted to nonhuman animals and have furthermore just undergone trial by acid as they passed through the stomach, it is likely that most of them will be unable to compete effectively enough with the resident bacteria to stay in the colon. Rather, most of them will probably be excreted without colonizing the gut.

If the incoming bacteria pass right through the colon and are excreted, how could they possibly cause us any problems? The answer is that during the day or two that they spend passing through the human colon, they could transfer their antibiotic resistance genes to native human colonic bacteria via HGT. After all, transfer of DNA by conjugation takes at most a few hours. Moreover, concentrations of the native bacteria are high and many of them are located on the surfaces of plant or other undigested food particles or the mucin that lubricates the intestinal lining. This is the sort of close proximity that facilitates conjugation.

As discussed in an earlier section of this chapter, although colonic bacteria are innocuous if they stay in the colon, when they escape from the colon during surgery or other abdominal trauma, they can cause deadly infections. Having a highly antibiotic-resistant normal bacterial population makes a person particularly vulnerable if that person ever has the misfortune to contract a postsurgical infection, most of which are caused by members of the patient's own microbiota or the microbiota of health care workers.

How likely is it that bacteria living in the human colon would acquire resistance genes from bacteria that are only passing through? Asked in a different way, how much HGT occurs naturally among human colonic bacteria or between them and bacteria that do not permanently inhabit the site? Initially, scientists assumed that the probability of such transfers was low, because in the laboratory the frequency of resistance gene transfer by naturally occurring gene transfer elements is usually quite low. For example, if an antibiotic-resistant colon bacterium (the gene donor) is mixed with a susceptible bacterium (the recipient) under optimized conditions, fewer than 1 in 10,000 of the bacteria that are potential recipients will receive the transferred gene. One would think, then, that transfer would be quite rare in the environment, where conditions would presumably be less favorable. As the old saying goes, though, truth can be stranger than fiction.

There is now substantial evidence that bacterial conjugation rates among gut bacteria are quite high, especially within the mucus layer. Moreover, intestinal microbiota can serve as major reservoirs of antibiotic resistance genes, particularly those belonging to Gram-negative bacteria within the large family called *Enterobacteriaceae*, including *Escherichia coli* and *K. pneumoniae*, and to the *Bacteroides* genus, a group of Gram-negative bacteria that accounts for about one-fourth of the bacteria in the human colon. In one study exploring the extent of HGT that occurs in the colon, the researchers looked at two sets of *Bacteroides* strains. One set of strains was isolated before 1970, while a second set of strains had been isolated in the late 1990s. The research team found that, while only about 20% of the pre-1970s strains carried a tetracycline resistance gene called *tetQ*, over 80% of the strains isolated in the 1990s carried *tetQ*, with the *tetQ* genes in all the strains from both sets having virtually identical DNA sequences. In contrast, the sequences of the genomes of the different *Bacteroides* species involved in this study differed by a much greater margin. This finding of identity in DNA sequences among *tetQ* genes in very diverse species of bacteria supported the hypothesis that the *tetQ* gene was transferred from one strain to another via HGT and did not evolve independently. This hypothesis was strengthened by the finding that the *tetQ* gene was indeed carried on a conjugally transferred integrated DNA element, called a conjugative transposon.

Aside from the finding that there has been a lot of conjugating going on in the colon over the intervening decades, there were two other surprising findings from this study. First, the fact that 80% of *Bacteroides* strains were carrying the *tetQ* gene, even though many of the strains had been isolated from healthy people with no recent use of antibiotics, indicated that carriage of this resistance gene was very stable (long-lasting even in the absence of selection pressure). This finding flew in the face of a favorite assumption on the part of physicians and many scientists that if you just stop using an antibiotic, the resistant strains will disappear because they are not as fit as the susceptible strains in the absence of the antibiotic. Obviously, *tetQ* and the transmissible element that carried it were not taking enough of a fitness toll on these *Bacteroides* strains.

A second unexpected finding from the study was that already in the pre-1970 period, when tetracycline had only begun to be used in clinical medicine, such a high percentage of strains (20% or more) already carried the *tetQ* gene. Where did this resistance gene come from, and what selection pressures led to it already being widespread before tetracycline use became intense in human medicine? This was not the first observation suggesting that resistance genes were around before antibiotics were used widely in medicine. The implication is obvious. There must be selection pressures other than clinical use that promote carriage of antibiotic resistance genes. The standard explanation for this phenomenon is that many soil bacteria produce antibiotics such as tetracycline or structurally like tetracycline and that contact with these antibiotics in the soil has caused bacteria found in the environment to become resistant to them even before they were harnessed by human scientists. The only problem with this theory is that the concentrations of antibiotics in soil are nearly undetectable, certainly not high enough to explain such a high level of resistance gene evolution seen in natural isolates.

Yet another theory is that antibiotics were originally signaling molecules that allowed bacteria to communicate with each other. In this view of antibiotics, the genes to which the antibiotics bound, which ultimately became resistance genes, were the receptors for these signals. There is no direct evidence for this theory except that in a few cases, antibiotics have been found to serve as inducers of gene expression in bacteria. For instance, the conjugative transposons that have been moving the *tetQ* gene around in the human colon are stimulated to transfer DNA by none other than the antibiotic tetracycline itself. Antibiotics not only select for survival of antibiotic-resistant bacteria; they may also, in some cases, stimulate the transfer of the resistance gene in the first place. Thus, widespread use of tetracycline by physicians (and possibly farmers) could have been responsible for stimulating the transfer of the *tetQ* gene among bacteria in the gut.

The study just described supports the contention that horizontal transfer of antibiotic resistance genes can readily occur in the colon, but it does not address the question raised earlier about gene transfer between swallowed bacteria transiently passing through and other members of the colonic bacterial population. The scientists conducting this study also noticed that a gene that conferred resistance to the antibiotic erythromycin (*ermB*) had appeared in *Bacteroides* species only in the post-1970s period. This gene had previously been seen only in Gram-positive bacteria such as *Staphylococcus, Streptococcus,* and *Bacillus* species. Sure enough, the *ermB* genes now being found in *Bacteroides* species were carried on conjugative transposons and were nearly identical at the DNA sequence level to the genes that had been found previously in Gram-positive bacteria. This result strongly supports the hypothesis that these genes were being transferred between different genera. More to the point, since *Streptococcus* and *Staphylococcus* species do not colonize the colon but merely pass through, finding identical genes in these genera and in *Bacteroides* species indicates that there is genetic communication between transient bacteria and the resident bacteria in the human colon.

What practical use are we to make of the type of information that has been described in this and the previous sections? First, we have learned that in some cases antibiotics such as tetracycline can stimulate bacteria to transfer antibiotic resistance genes. Thus, when a new antibiotic is being evaluated for possible use, more than its predilection for selecting for bacteria that have become resistant needs to be considered. The effect of the antibiotic on stimulating transfer of resistance genes also needs to be evaluated, insofar as we know enough about various resistance gene transfer elements to do that. Second, it is no longer safe to assume that, once evolved, an antibiotic resistance gene will stay within a single species and in a single location. This observation underscores the importance of limiting the use of antibiotics as much as possible to cases in which they are essential for maintaining health.

Antibiotics as environmental pollutants

A continuation of the story above about vancomycin resistance brings us back to the agricultural use of antibiotics that was introduced in chapter 3. In the European Union, although not in the United States, an antibiotic called avoparcin was approved in the 1990s for use as a growth promoter. Have you ever heard of avoparcin? Neither had most people involved in the monitoring of resistance patterns in humans. Avoparcin is an analog of vancomycin, and it selects for resistance to vancomycin as well as to itself. Not surprisingly, after a couple of years of avoparcin use in Europe, vancomycin-resistant strains of *E. faecium* began to appear in farm animals. More troubling was the fact that vancomycin-resistant *E. faecium* strains also began to be found in the intestines of urban European adults.

The same vancomycin-resistant enterococci were not found in people in the United States, where avoparcin had not been approved for use in agriculture. Instead, extensive use of vancomycin in hospitals in the United States had spawned different types of vancomycin-resistant *E. faecium* strains, and these seemed to be limited to hospitals and did not move into the community. Europeans, by contrast, had not had much of a problem with vancomycin-resistant enterococci in their hospitals at first because vancomycin use was much more severely limited in human medicine than was the case in the United States. The different ecologies of the vancomycin-resistant strains in Europe and the United States supported the hypothesis that the European urban adults with vancomycin-resistant strains had acquired them through the food supply from farms that used avoparcin as a growth promoter. Consistent with this hypothesis, when the European Union abruptly banned avoparcin use in agriculture after being alerted to its dangers, the carriage of vancomycin-resistant enterococci in the colons of urban adults decreased quickly from 10% to 4%. This experience, which was followed carefully by European scientists, is one of the best examples of a possible linkage between farm use of antibiotics and antibiotic resistance patterns in human intestinal bacteria.

There is now overwhelming evidence linking the use of antibiotics in human medicine and agricultural animals to contamination of drinking water sources, municipal wastewater, and crop fields with these antibiotics. Naturally, a corresponding increase in antibiotic resistance among environmental bacteria present in these areas is also observed. In addition, it appears that at least some resistance genes are moving quite widely among certain bacterial species found in natural settings, including species normally found in very different environments.

Many antibiotics consumed by patients are quite stable and are subsequently excreted from the body and flushed down the toilet. Significant amounts of antibiotics and their metabolites or degradation products (antibiotic residues) have been reported in soil, wastewater, and sewage systems. Veterinary consumption and agriculture use as growth promoters likewise result in discharge of antibiotics in manure and soil environments and is associated with groundwater and drainage runoff or with water used for crop irrigation. Until recently, environmental contamination with antibiotic discharge into wastewater effluents by pharmaceutical companies was also a serious problem. Fortunately, several prominent pharmaceutical companies have voluntarily taken measures to reduce the amount of antibiotics and antibiotic residues released by their manufacturing processes into wastewater systems.

The overall impact of antibiotic release into the environment is still not fully known, but one thing is clear: the above practices place a persistent selective pressure on the bacteria in these environments both to maintain antibiotic resistance and to share resistance genes with other bacteria present. Other potential effects could also be imposed on aquatic and soil ecosystems, where the composition of the normal microbiota could be perturbed by the presence of the antibiotics. What this might do to the plants, insects, and animals in that ecosystem is not clear yet, but there is accumulating evidence that suggests the outcome is not good. Another troubling aspect of the rise in antibiotic levels in the environment from human or animal usage is that their presence indicates that they are clearly not being effectively removed during conventional water or wastewater treatment processes.

Fluoroquinolones: a widely (too widely?) used family of antibiotics

In the fall of 2001, federal office workers who had worked in buildings that received anthrax-laced letters lined up to receive Cipro, the trade name of the fluoroquinolone antibiotic ciprofloxacin. Other uses of ciprofloxacin include prevention of bacterial infections in cancer chemotherapy patients, treatment of postsurgical infections, and treatment of urinary tract infections. Cipro and other fluoroquinolones are among the most frequently prescribed antibiotics today. These drugs are considered an important part of the antibiotic armamentarium and are currently regarded as one of the front-line classes of antibiotics to defend patients against infections caused by bacteria that are resistant to other antibiotics. Fluoroquinolones are now being used to treat tuberculosis (TB), a very dangerous disease, especially in cases where the bacterium that causes it has become resistant to the traditional anti-TB drugs. Overuse of this very important type of antibiotic is already giving rise to resistance to the fluoroquinolones, an alarming development, to say the least.

Fluoroquinolones like Cipro are chemically very stable and not easily degraded under even extreme environmental conditions such as heat and acidity. Fluoroquinolones are excreted unchanged from the body in urine and are then discharged into hospital or municipal wastewater or sewage systems, where they then can absorb onto particles in sludge, which sometimes is used as a fertilizer in agriculture. This results in continuous introduction of a bioactive compound into the environment that then persists. Some of the levels of these antibiotics found in aquatic systems appear to have resulted in resistant strains of the pathogen *Salmonella* that have had deleterious outcomes on some aquatic animals and insects.

The widespread use, some would say abuse, of fluoroquinolones is not limited to human medicine. In 1996, the U.S. Food and Drug Administration (FDA) approved the use of the fluoroquinolone enrofloxacin by chicken farmers, who administer it to chickens in their water to prevent infections caused by *E. coli*. Such infections can wipe out an entire flock or at least slow their growth, which is almost as bad an outcome. Thus, farmers want to treat an entire flock if even a single bird is diagnosed with this type of intestinal disease or if the disease is present in the area.

When the decision to approve enrofloxacin was made, concern about a rise in resistance to fluoroquinolones was not nearly as high as it is today, and the farmers' organizations made a persuasive case, based on scientific evidence available at the time, that the use of fluoroquinolones in agriculture was safe. Subsequently, scientists at the Centers for Disease Control and Prevention (CDC) have documented a rise in fluoroquinolone resistance in *Salmonella enterica*, a common cause of foodborne infection. This rise in resistance began at about the same time the antibiotic was introduced into agricultural use.

Of course, this increase in resistance also started about the same time that fluoroquinolones began to be used widely to treat human disease as well, so the contention that agricultural use of fluoroquinolones is primarily responsible for the rise in the number of resistant *Salmonella* strains is, understandably, controversial. How would using enrofloxacin to prevent an *E. coli* infection in chickens affect *Salmonella* strains that might infect humans? *Salmonella*, the cause of salmonellosis, also colonizes chickens, although it does not cause disease in chickens the way it does in humans. *Salmonella* would thus be exposed to the same antibiotic used to prevent another type of infection in the chickens. And, of course, any resistance against one antibiotic would also work against the other. The complication arises when one considers that because salmonellosis is a self-limiting infection, antibiotics are seldom used to treat patients with it. However, in some cases, granted just a small minority of cases in an outbreak, the infection can move from the intestinal tract into the bloodstream. Such a dispersed infection can lead to lethal sepsis. Children and the elderly are the most likely to develop the bloodstream form of the infection. In this case, it would be critical to have an effective antibiotic such as Cipro to use against *Salmonella*, but with the use of enrofloxacin, the likelihood that the *Salmonella* might be resistant is higher.

In some ways, the emphasis on fluoroquinolone resistance of *Salmonella* strains distracts attention from a potentially more serious problem, a rise in the resistance of other bacteria such as *Enterococcus* species that are already resistant to many other antibiotics and can cause life-threatening postsurgical infections in humans. The concern is that if fluoroquinolone-resistant *Enterococcus* species found in chickens can colonize humans and the bacteria are then ingested in food, they could ultimately cause serious postsurgical infections. Where do we draw the line on the agricultural use of antibiotics?

Those who want absolutely no risk to human health advocate a ban on enrofloxacin use by chicken farmers, a move that would probably result in an increase in the price of chicken for consumers and would be hard on farmers. Those who are willing to tolerate a manageable level of risk might focus on the benefits to farmers of this use of antibiotics and ask what its true risk is to the treatment of human infections. Clearly, people are not currently dying by the thousands because of fluoroquinolone-resistant salmonellosis. This picture of fluoroquinolone resistance may change, however, if agricultural use of fluoroquinolones gives rise to resistant human pathogens that cause postsurgical infections or infections in cancer patients. Will fluoroquinolone-resistant enterococci and other human pathogens for which antibiotic treatment is an absolute necessity begin to take a greater toll? The answer to this question is a definite yes, since fluoroquinolone-resistant pathogens already are among us and already have begun to cause medical problems. The question is whose fluoroquinolone use is to blame: that of physicians, that of farmers, or both?

Food safety—contribution of GM foods and globalization to resistance

Given that many bacteria can transfer antibiotic resistance genes among one another, and some can also acquire DNA from the environment through uptake mechanisms, many folks began to wonder about the risks posed by consumption of genetically modified (GM) foods containing antibiotic resistance genes. Other folks worried about the potential for rapid spread of the antibiotic-resistant bacteria and antibiotic resistance genes through the globalization of the food industry. In each case, the overriding question was: Can eating GM food or imported food put me at risk of getting infections that are resistant to antibiotics?

The flap over GM foods—lessons from the gene transfer front. So far in this chapter, conjugation has been the star of the DNA transfer show, but transfer of antibiotic resistance genes can also occur by transformation, the direct uptake of DNA from the environment. This well-known fact led opponents of GM foods to raise antibiotic resistance as a possible safety concern associated with GM foods. Until the advent of recent gene editing technologies, GM plants usually carried an antibiotic resistance gene, usually either an ampicillin resistance gene or an aminoglycoside resistance gene, because these genes were often employed as selectable markers on plasmids used to clone the genes that were destined to be moved into the plant cells.

For instance, when the plasmid carrying the insecticidal Bt toxin gene was moved into the plant cell to generate a Bt crop, the antibiotic resistance gene went along for the ride and was also integrated into the plant genome. The antibiotic resistance gene that enters the plant genome has no effect on the plant. Not only is it a bacterial gene that is not expressed in the plant, but plant cells, being eukaryotic cells, are naturally resistant to antibiotics such as ampicillin. Plants, of course, do not have cell walls like bacteria and thus are impervious to peptidoglycan synthesis inhibitors.

The concern expressed by the opponents of GM crop biotechnology was that the antibiotic resistance gene, such as the gene encoding beta-lactamase, lodged in the plant genome might be released from the plant cells during digestion in the human intestinal tract, and the released DNA taken up by human intestinal bacteria. Should the beta-lactamase gene be retained by the transformed intestinal bacteria, then these bacteria might become resistant to the beta-lactam ampicillin. Numerous groups of scientists met all over the world to discuss this possibility. Regulatory agencies in Europe, the United States, and many other countries consulted antibiotic resistance experts to assess the possible magnitude of this danger. Almost unanimously, the microbiologists who were consulted concluded that there was little or no danger of such a gene transfer event happening.

For one thing, only a small number of bacteria are capable of spontaneously taking up DNA from the environment. Therefore, laboratory scientists who want to introduce DNA into bacteria like *E. coli* need to electrocute these bacteria or chemically force them to artificially take up DNA from the medium outside the cell. If the DNA is taken up, it has to integrate into the chromosome to be retained. That can only be done through homologous recombination, a process that requires portions of the incoming DNA to be identical to DNA in the recipient bacterium's chromosome. The effect of this is that most gene transfers that occur in nature by transformation occur between very closely related strains of the same species. That is, an antibiotic resistance gene derived from *E. coli* or closely related bacterial species, as most of the marker genes on cloning vectors were, could not enter and survive in unrelated bacteria.

Despite these compelling arguments that the movement of an ampicillin resistance gene from a plant cell to an intestinal bacterium was extremely unlikely, the argument finally boiled down to the question of what the consequences would be if such an event did occur. The answer to this was easy: not much of anything serious would happen. Physicians already have much more advanced beta-lactam antibiotics that can counter any ampicillin resistance gene (more details to come in chapter 8). Therefore, if the genetic engineers creating the GM foods are mindful of the antibiotic resistance gene used (i.e., they do not use one that could confer resistance to a clinically important antibiotic in current use), even if a gut bacterium acquires a resistance gene from the GM food, the outcome would be inconsequential.

Globalization of the food industry—resistance spread on a massive scale. Chances are that many of the foods you typically eat originate from another country. In fact, most foods that we commonly eat are products of global trade not just with other states across our own country but also with countries from around the world. In other words, over the past few decades the food industry has expanded to encompass all parts of the world. This globalization of the food industry means that many food commodities, such as coffee, bananas, corn, and beef, grown in one country are being exported to another country where they are consumed. Global commerce agreements have lowered barriers to free trade, and modern refrigeration, preservation, and transportation methods have enabled the rapid transporting of large quantities of food around the world. While the economic, nutritional, and cultural pros and cons of food globalization can be debated, there is no doubt about the significant increase in food safety threats associated with this large-scale industrialization of food production, processing, and distribution.

This globalization of the food industry also means that foodborne illnesses are no longer associated with eating contaminated potato salad at a Sunday picnic social, poorly canned green beans at the family dinner, or meals where someone forgot to properly wash their hands. From a food safety perspective, antibiotic resistance poses an ever-increasing health and socioeconomic risk to exposed workers and consumers around the world. Globalization of trade in food animals, produce, crops, and other foodstuffs means that resistant bacteria emerging in one geographical location or from one source can readily spread to other bacteria in other sources and in other locations. This spread can occur rapidly at any stage along the process from production and harvesting on the farm to processing and distribution centers to food handlers and end consumers, and importantly, it can affect all participating parties, regardless of country of origin or destination.

The problem of emergence and spread of antibiotic resistance in animal food production is compounded by the increased potential for transfer of the antibiotic resistance genes to the resident bacteria in the intestines of consumers. In both cases, as mentioned before, the resistant bacteria can then be transmitted to other hosts or released into the environment through waste effluents. Many of the foodborne bacteria bearing antibiotic resistance genes are also notorious for their virulence properties, and many clinical isolates of these MDR bacteria fall into the categories of priority pathogens or superbugs mentioned in chapter 3.

Salmonella and *Campylobacter* species are particularly problematic for the poultry industry, causing large foodborne outbreaks of fever and diarrhea around the world due to contaminated meat and eggs. Many clinical isolates of *Salmonella* and *Campylobacter* associated with severe illness show MDR to several important front-line antibiotics like beta-lactams, tetracyclines, and fluoroquinolones. *Enterococcus* species are part of the normal microbiota of both humans and animals, and as such, often serve as reservoirs of antibiotic resistance genes. Antibiotic-resistant, pathogenic *E. coli* strains harboring potent toxins have been associated with deadly foodborne gastrointestinal illnesses caused by contaminated raw or undercooked ground beef and poultry meat products; fresh raw produce such as lettuce, sprouts, and spinach; and unpasteurized milk or milk products. The striking fact about these outbreaks is that they often involve cases in multiple states or multiple countries, all caused by the same antibiotic-resistant pathogen.

And, because the contaminated food was so rapidly transported to so many different destinations, it was already quite widespread before health authorities realized the problem and could put in place containment measures to halt the spread.

Regulatory decisions that impact the spread of antibiotic resistance

The bioterrorism attack that started in October 2001, when an unknown person mailed letters laced with spores of *Bacillus anthracis*, started a panic about the safety of the postal system, but also revealed a regulatory quandary that has spurred a lot of public discussion of the importance of antibiotics and the ways in which different antibiotics are regulated. After several cases and deaths due to anthrax, caused by the spore-forming bacterium *B. anthracis*, first appeared, widespread panic ensued when the vehicle for dissemination of the spores was discovered unexpectedly to be letters delivered by the postal system. Fortunately, scientists already knew a lot about anthrax and its treatment. Antibiotics given soon after exposure will prevent the most fatal form of anthrax infection, inhalation anthrax. Cipro was administered to workers in the congressional office buildings and news buildings where letters containing the spores had been opened. By contrast, an antibiotic called doxycycline was administered to postal workers who might have been exposed to the spore-laden letters. In both cases, the timely intervention of antibiotics prevented further deaths.

Soon, people all over the United States were asking their physicians whether they should take Cipro. The course of Cipro therapy required taking Cipro daily for at least 60 days, an unusually long period of therapy for an antibiotic. This course was chosen because no one knew how long Cipro, which was a relatively new drug at the time, would take to clear the bacteria from the body. There had been no cases of human anthrax in the United States for many years, so there was no standard of reference. As it turned out, few people completed the full course of Cipro therapy because of the unpleasant side effects (mainly nausea and diarrhea).

Many other antibiotics besides Cipro are effective against anthrax. One is doxycycline, an antibiotic that is a lot cheaper and has fewer side effects than Cipro, but many people felt, wrongly, that Cipro was the best choice. A heated debate over the relative merits of Cipro and doxycycline developed early in the anthrax scare period. One reason, already mentioned, was that congressional staff members and other elite groups were given Cipro, whereas postal workers were given the much less expensive doxycycline. Reporters, postal workers, and their unions immediately sensed a class disparity issue. Ironically, the supposedly favored congressional workers were given a drug that was quite expensive initially (up to $700 for the 60-day course of therapy) and had unpleasant side effects, whereas the postal workers received an equally effective but much cheaper antibiotic with fewer side effects.

What might have led to the disparate use of these two drugs? Cipro was, technically, the only drug approved by the FDA for the treatment of anthrax at the time.

Doxycycline was not approved, in the sense that anthrax was not listed among the diseases for which doxycycline was recommended. This instance illustrates a problem in interpreting the meaning of "FDA approval" of a drug. There are many examples in which the FDA has not approved a drug for a certain application because it was ineffective or dangerous, but this was not the case for doxycycline. Instead, economics and the timing of approval were responsible for the "approved" status of Cipro and the "not approved" status of doxycycline. Ciprofloxacin was a relatively new drug that was still under patent protection, hence, its high price tag. When the company that produced Cipro sought FDA approval for this drug, it included a long list of possible applications of the new antibiotic. Treatment of inhalation anthrax was included on this list, despite no known human cases of inhalation anthrax in the United States for decades.

When doxycycline came up for FDA approval in the 1960s, no one even considered listing anthrax as a disease treatable by the antibiotic because anthrax was so rare and there was a strong treaty in force that banned the development and use of biological weapons. Even after there was talk of the possible use of *B. anthracis* as a bioterror weapon in the 1990s during the first Gulf War, it was hardly worthwhile for the manufacturers of doxycycline to get an FDA approval because doxycycline was no longer under patent protection. So, although infectious disease experts agreed that doxycycline would be an effective treatment for anthrax in this case, the fine print on the literature about the drug did not include this application.

As an outcome of the anthrax incident and the ensuing debate regarding large-scale use of unapproved drugs, the Emergency Use Authorization (EUA) program was established in 2004, which enables the FDA to authorize use of an unapproved medication during any public health emergency that is declared by the Department of Health and Human Services, the Department of Homeland Security, or the Department of Defense. Under the EUA program, the FDA can issue emergency dispensing orders without requiring an individual prescription for each recipient of the drug, and it allows for waiver of Current Good Manufacturing Practice (CGMP) requirements (regarding handling, storage, and product quality) when appropriate. In recent years, the FDA has issued numerous EUAs for critical, urgently needed experimental drugs, diagnostics, and vaccines during the Ebola virus, Zika virus, and severe acute respiratory syndrome coronavirus (SARS-CoV-2; COVID-19) pandemics.

So, you might ask, how did this regulatory issue impact the spread of antibiotic resistance? Unfortunately, resistance to fluoroquinolones develops relatively easily since a single mutation in the DNA gyrase enzyme can make a bacterium completely resistant. For this type of resistance, the speed with which resistant strains appear is proportional to the amount of the antibiotic used. Because Cipro was dispensed so widely to so many people, it is perhaps not surprising that fluoroquinolones have gone from being reserved for treatment of the most serious human infections to possibly becoming the class of antibiotic that was the most rapidly lost as a treatment option.

POINTS TO PONDER

To what extent do farmers have a vote in debates about antibiotics like tylosin and virginamycin (a streptogramin antibiotic also used as a growth-promoter in animals)? They are a small minority of the population in terms of actual votes in an election, but they are absolutely critical to our survival. How do we weigh the need of farmers to make a decent living against the possibility that some of their practices might, but have yet to be proven to, increase the speed with which antibiotic-resistant human pathogens emerge?

The effective use of antibiotics such as ciprofloxacin (Cipro) and tetracycline to prevent or cure anthrax was reassuring in the bioterrorism incident of 2001, since the *B. anthracis* strain used was still sensitive to those two different classes of antibiotics. However, another terror-inducing concern quickly came to light that caused some turmoil in the eyes of those responsible for defending our well-being. Some folks immediately began to wonder if strains of *B. anthracis* resistant to multiple classes of antibiotics could be the next bioterror threat. After all, it was just demonstrated that some individuals were fully capable of harming others with deliberate release of a deadly pathogen, and many scientists openly confirmed that introducing antibiotic resistance genes into *B. anthracis* would be relatively easy to do, even in a makeshift, low-budget laboratory. Some folks further speculated that antibiotic resistance genes could also be introduced to other deadly pathogens that could be used by bioterrorists, thereby generating new additions to the list of biothreat agents, collectively called "select agents" by the CDC.

Consequently, combatting select agents became and remains a major global health priority in the United States and elsewhere, with the United States alone having already spent many billions of dollars toward this effort. Clearly, this incident brought the possibility of deliberate deployment of biothreats as agents of terror into the realm of reality. So, with what you know about bacterial behavior and antibiotic resistance, could the scenario above get worse? Which scenario is scarier, the human-made one described here or the one that is already happening naturally in the environment and in hospitals around the world? Is this something that we need to worry about? What do you think—could addressing one problem also serve to address the other?

8 The Escalating Crisis in Antibiotic Availability Amidst Renewed Efforts

There are a few obvious approaches to dealing with the antibiotic resistance problem. First, educational programs already under way can advise physicians, patients, and farmers about the appropriate use of antibiotics. While such efforts have already resulted in positive changes in the behaviors of each group, this type of effort alone is unlikely to bring about the societal-level course correction that may be necessary to preserve antibiotics or to lead to effective new ones. After all, we already found out that once antibiotic resistance is here, it does not simply go away.

A second response would be to forcefully limit the use of antibiotics. Laws have been put in place to restrict agricultural applications of antibiotics to prophylactic use only, preventing their use as growth promoters. As of 2017, the U.S. Food and Drug Administration (FDA) banned antibiotic use for growth-promoting purposes in the United States. However, in the large-scale animal food industry, there may not be much of a difference between growth promotion and disease prevention when it comes to actual industrial practices, especially considering that there is no limit placed on how long an antibiotic can be administered and to how many animals at a time. For example, to prevent liver abscesses, an issue that can cause significant economic loss if not addressed, farmers still routinely give daily doses of the macrolide antibiotic tylosin to cattle in feedlots. The issue here is that some feedlots can hold as many as 100,000 cattle. At this scale, even short courses of antibiotic treatment can lead to release of antibiotic-containing effluent

Revenge of the Microbes: How Bacterial Resistance Is Undermining the Antibiotic Miracle, Second Edition.
Brenda A. Wilson and Brian T. Ho.

in the environment and to increased macrolide resistance in gastrointestinal bacteria, which as discussed in chapters 3 and 7 contributes to the spread of antibiotic resistance.

In addition to enforcing antibiotic use limitations in agriculture, similar restrictions can be applied to clinical uses. A physician's freedom to prescribe antibiotics for inappropriate applications could be curbed by placing veto power in the hands of pharmacists or infectious disease specialists. This kind of restriction would not be unprecedented—after all, there are limitations on a physician's rights to prescribe such drugs as morphine. In fact, some form of oversight over antibiotic prescriptions has already been introduced in a few large urban hospitals. For example, a physician deemed to be misprescribing or overprescribing antibiotics may receive an admonitory letter from the hospital infection prevention and control officer. Every emergency response employer in the United States is required to have a designated infection prevention and control officer, who watches for signs of increases in hospital-acquired infections, which can arise from bad hygienic practices or overuse or misuse of antibiotics. While this type of arrangement may work in a large hospital, it has the potential to turn into a bureaucratic nightmare if extended to physicians practicing in the community, so further efforts will need to be made to come up with rational solutions that can work at all levels.

The final way to deal with increased antibiotic resistance would be simply to step up the discovery of new antibiotics. This approach was the main avenue taken for nearly 50 years after the first antibiotics were introduced. After waning for a couple of decades, it has been revitalized by a number of technological developments, including high-throughput screening, chemical synthesis innovations, recombinant protein engineering, computational technologies, and advances in large-scale comparative genomic sequencing. Many people will be surprised to hear that this obvious technological fix is no longer being aggressively pursued. If anything, the movement in large pharmaceutical companies has largely been in the direction of reducing research and development of new antibiotics, with many shutting down or cutting back on their antibiotic discovery programs. Nearly all the big pharmaceutical companies have cut back on or eliminated their in-house programs for antibiotic discovery, leaving only a few remaining companies with active antibiotic programs. When the antibiotic discovery task was switched to smaller biotechnology companies, many of them also caved under the financial burden of low return on investment, with only a few being successful enough to be purchased by some of the remaining larger companies. Nevertheless, there have been some spurts of effort with some success, and so we thought it might be useful to provide some historical and economic context about those advancements from the pharmaceutical industry perspective, to understand why things have progressed to where they are today.

ESCALATING BACTERIAL RESISTANCE TO ANTIBIOTICS SPAWNS NEW GENERATIONS OF ANTIBIOTICS

Since the earliest antibiotics were discovered and introduced into clinical use, the pharmaceutical industry has continued to turn out new versions of old favorites as well as new types of antibiotics. In some cases, these new versions are designed to have improved pharmaceutical properties, such as better distribution in the body and fewer side effects. In most cases, however, the goal has been to modify the antibiotics to counter the latest version of bacterial resistance strategies.

New versions of antibiotics are often described as if they were generations of a family. Thus, penicillin and other early beta-lactam antibiotics are called first-generation beta-lactam antibiotics, whereas later alterations of the basic penicillin structure have been called second generation, third generation, and so forth. The multiplication of drug generations is not good news for people who will contract serious infections, because it bears testimony to the continued ability of bacteria to become resistant to each new generation of antibiotic. Also, each new generation bears a higher price tag because the newer antibiotics are still under patent protection and still have cost-recovery prices associated with them.

The fact that we are currently in the fourth, fifth, or even later generation of most antibiotics, after just 70 years of antibiotic use, should be warning enough that the race between us and microbes is neck and neck. Until recently, the emphasis on maintaining antibiotic efficacy has focused on creating yet newer generations of antibiotics. However, scientists and physicians are now beginning to realize that another approach is needed—finding ways to prevent the development of bacterial resistance. Options include more prudent use of antibiotics to reduce the selection pressures that encourage resistant bacteria to emerge and persist. Another strategy is directly targeting and inactivating bacterial mechanisms for resisting antibiotics.

Scientists strike back: combatting beta-lactamases

To illustrate how the battle between antibiotic discovery by researchers and gain of antibiotic resistance by bacteria progresses over time, consider the beta-lactam family of antibiotics. Shortly after the introduction of penicillin, a bacterium producing a beta-lactamase, which destroys the beta-lactam ring and inactivates the antibiotic, emerges and the gene encoding this beta-lactamase spreads to other bacteria. Score one for the bacteria, but scientists have some tricks up their sleeves too. The scientists respond to combat penicillin-inactivating beta-lactamases by developing modified forms of penicillin that are no longer attacked effectively by these beta-lactamases. Note that in Fig. 5.4 the beta-lactam ring is surrounded by other chemical groups. Some of these groups participate in binding the antibiotic to the beta-lactamase, which enables the enzyme to hydrolyze and inactivate the antibiotic (see Fig. 6.2).

If the groups are altered, the ability of the beta-lactamase to bind to and inactivate the antibiotic is decreased. These new second-generation antibiotics have been called beta-lactamase-resistant beta-lactams.

Unfortunately, as rapidly as the scientists produce new forms of beta-lactamase-resistant beta-lactam antibiotics, the beta-lactamases themselves mutate, adjusting their structures so that they become able to bind and degrade the second-generation antibiotics. These mutant enzymes are called extended-spectrum beta-lactamases (ESBLs) because they now have an extended spectrum of new beta-lactams that they can act on. Unfortunately, as soon as the genes encoding ESBLs are generated, they then spread rapidly among bacteria around the world, thus setting off another round of new third-generation beta-lactam antibiotics to counter the ESBLs. And so the cycle continues.

Scientists subsequently invented a second way to deal with bacterial beta-lactamases, by developing an antibiotic preparation that contains both the antibiotic and an inhibitor of the beta-lactamase that attacks it. A widely used example of this type of preparation is a combination of the beta-lactam antibiotic amoxicillin and a beta-lactamase inhibitor, clavulanic acid. This mixture is marketed under the trade name Augmentin, a popular drug for treating many kinds of infections. The antibiotic component of Augmentin is amoxicillin (a member of the penicillin family), which was falling out of favor because bacteria were becoming increasingly resistant to it by producing an amoxicillin-inactivating beta-lactamase. To counter the action of this beta-lactamase, Augmentin also contains clavulanic acid (also a beta-lactam compound) that binds and inhibits the bacterial enzyme so that it can no longer destroy amoxicillin, leaving the amoxicillin to carry out its lethal task. Interestingly, clavulanic acid has no antibacterial activity itself. Its sole role is to bind to and inactivate the beta-lactamases that are trying to destroy amoxicillin. Clavulanic acid molecules, like members of a football defense line, keep the beta-lactamase out of action so that the offensive line amoxicillin can score a "touchdown" by weakening the bacterial cell wall.

Bacteria develop another resistance strategy that trumps the beta-lactamase inhibitors

The early successes of the antibiotic plus beta-lactamase inhibitor combination strategy made scientists optimistic that they were finally winning the fight against bacteria that were becoming resistant to beta-lactam antibiotics. All they had to do was keep improving the ability of the beta-lactamase inhibitors to inactivate the ever-evolving bacterial beta-lactamases, in essence generating second- and third-generation inhibitors to go along with the next generation of beta-lactams. In this way, old standbys like amoxicillin and ampicillin could be recycled.

The Gram-positive bacteria, and subsequently the Gram-negative bacteria, had an unpleasant surprise in store for these scientists, however.

One resistance mechanism that is not countered by beta-lactamase inhibitors is mutations that occur in the penicillin-binding proteins, which are the targets of the beta-lactam antibiotics, because no beta-lactamase is involved in this type of resistance. A bacterium can, of course, combine mutant penicillin-binding proteins with beta-lactamases for a very effective one-two punch, but at best the beta-lactamase inhibitors will only counter one of the resistance mechanisms. Score one more for the bacteria.

REDISCOVERING/REPURPOSING OLD OR ABANDONED ANTIBIOTICS

The growing medical need for new treatment options when the latest versions of antibiotics no longer work and new ones are getting scarce has led some researchers and large pharmaceutical companies to reconsider older antibiotics that had previously fallen by the wayside due to various undesirable properties or costs associated with their use or development. The strategy of repurposing such drugs or drug candidates is gaining momentum in the industry. The advantages of repurposing become obvious when you consider that most of these drugs or drug candidates have already advanced in the approval process since a lot of the legwork for the early discovery, lead optimization, formulation, and safety testing has already been done in previous studies. Thus, the costs associated with preclinical development of the drug can be greatly reduced and the approval process timeline accelerated.

Another type of cell wall synthesis inhibitor: vancomycin

The success of the beta-lactam antibiotics, both in terms of efficacy and safety, was so impressive that little attention was paid for a long time to another antibiotic that also inhibited bacterial cell wall synthesis: vancomycin. Vancomycin and a related antibiotic called teicoplanin had been discovered in the 1960s. Vancomycin had two major drawbacks. First, the initial preparations had a brown color, a trait that caused some scientists to call vancomycin "Mississippi mud"—not much more appealing than the gramicidin "earwax" described in chapter 4. The brown color was a tip-off that this vancomycin preparation contained several contaminants that would make it not only less effective but also possibly less safe and more difficult and expensive to purify.

A second drawback was that vancomycin was only effective against Gram-positive bacteria. The reason for this is that vancomycin is a big, bulky molecule that cannot diffuse through the outer membrane porins of Gram-negative bacteria. When vancomycin was first discovered, physicians preferred to prescribe broad-spectrum antibiotics that were effective against all kinds of bacteria, Gram positive as well as Gram negative. This practice allowed them to dispense with the time-consuming process of identifying the bacteria through diagnostic tests. It also made it possible, in theory, to use a single antibiotic for many infectious disease applications, i.e., for diseases caused by different types of bacteria. Such broad-spectrum antibiotics are particularly important in the early critical period when a physician suspects that the patient has a bacterial infection but has no idea of its species identification or antibiotic susceptibility. It is likewise important when the infection has progressed to the point of life-threatening sepsis, where elimination of the bacteria in the bloodstream cannot wait for diagnostic tests to be performed.

Another development that made vancomycin unappealing was that during the 1960s and 1970s, the Gram-positive bacteria, which had dominated the disease picture in the period from 1940 to 1950, decreased rather dramatically in the number of cases they caused. Gram-negative bacteria such as *Escherichia coli*, *Klebsiella* species, and *Pseudomonas* species became the bacteria to beat in hospitals and clinics. In the 1980s, however, Gram-positive bacteria such as staphylococci and streptococci began to roar back into a position of prominence, for reasons that are still not completely clear. A likely explanation is that the use of antibiotics that were most effective against Gram-negative bacteria allowed the Gram-positive bacteria to reclaim their starring role in the infectious disease arena. Vancomycin, which had been collecting dust on the shelf prior to the 1980s, began to look pretty good after all. Scientists were able to obtain a purer preparation that had lost the contaminants responsible for the brown color of the earlier preparations, and vancomycin soon became a front-line antibiotic for treating those Gram-positive bacteria that had become resistant to other antibiotics.

Today, one of the hottest-growing areas of antibiotic discovery is finding more antibiotics that are effective against Gram-positive bacteria. Ignoring the Gram-negative bacteria, however, is as unwise today as was ignoring the Gram-positive bacteria during the period from 1960 to 1970, because it is now clear that the spectrum of disease-causing bacteria can fluctuate, and as we can see, both types of bacteria make up the list of priority and ESKAPE pathogens plaguing us today.

Vancomycin, like penicillin, inhibits the cross-linking reaction that finishes the construction of the peptidoglycan cell wall, but it does so in a different way. Whereas penicillin binds to and inactivates the enzymes that make the cross-links, vancomycin binds to the peptides that are slated to become part of the cross-linked peptidoglycan structure. Enzymes have a very precise lock-and-key interaction with their substrates, and anything that interferes with that interaction can stop the enzymes in their tracks. Vancomycin is a large, bulky molecule that binds to the part of the peptide that the cross-linking enzymes would normally interact with and thus physically prevents the enzymes from carrying out the cross-linking reaction.

How could a bacterium become resistant to such an antibiotic? One would have thought that vancomycin would have been the ultimate resistance-proof antibiotic because becoming resistant to it would require the bacteria to change their cross-link peptide structure, a change that not only might confuse the cross-linking enzymes but would require a new pathway for synthesis of the peptide. Also, the bacterium would have to get rid of the old peptide or it would remain susceptible to vancomycin's action. Unfortunately for us, some bacteria have managed to accomplish this feat.

Most of the resistance strategies described so far in this book involve a single resistance gene that encodes an enzyme that inactivates the antibiotic or an altered antibiotic target that is no longer affected by the antibiotic. Resistance to vancomycin, on the other hand, involves several genes encoding several proteins that comprise an entire pathway for changing the peptidoglycan cross-linking peptides into a form that no longer binds vancomycin but will still be cross-linked by bacterial enzymes. There is also a gene that encodes an enzyme that degrades the original terminal dipeptide (D-alanine-D-alanine) part of the cross-linking peptide.

There are some Gram-positive bacteria like *Lactobacillus* species that are naturally resistant to vancomycin because they have peptides in their peptidoglycan that do not have a terminal D-alanine-D-alanine dipeptide. Fortunately, this resistance seems to be an innate metabolic trait, a naturally occurring difference from other bacteria in the peptidoglycan peptide cross-link, and not a characteristic that can be transmitted to other bacteria. However, the most feared development is the emergence of strains of *Staphylococcus aureus* that start out being resistant to many antibiotics, including beta-lactam antibiotics, and have now become resistant to vancomycin. These strains have been called VRSA for vancomycin-resistant *S. aureus*. Fortunately, the VRSA strains isolated so far have proven to be susceptible to at least one other class of antibiotic, but since *S. aureus* has shown an impressive adeptness at acquiring resistance to many antibiotics, this situation may not last very long.

What will replace vancomycin and its relatives? Given that so many big pharmaceutical companies have abandoned the search for new antibiotics for economic reasons, the list of possible vancomycin replacements has become a very short one. This fact underscores the importance of protecting vancomycin and more recent derivatives of this antibiotic from overuse in the hope that the lifetime of these now-vital antibiotics can be prolonged until substitutes are found. This is going to be hard to do because surgeons, worried about postsurgical infections caused by bacteria that are resistant to other antibiotics, have increasingly favored vancomycin to prevent and treat Gram-positive infections. This type of decision increases further the overuse of vancomycin and provides a selective pressure for the emergence and dominance of VRSA and other Gram-positive bacteria.

It is difficult to do effective finger-pointing at physicians who may have a bit of an itchy finger on the vancomycin trigger when you consider that an analog of vancomycin, avoparcin, has been used as a livestock feed additive in Europe (see chapters 3 and 7). Avoparcin has a different name from vancomycin, but it cross-selects for resistance to vancomycin. Fortunately, the use of avoparcin in agriculture has been banned everywhere. Unfortunately, vancomycin resistance in *Enterococcus* and *S. aureus* continues to persist in the environment and in clinical settings. The avoparcin example illustrates dramatically the importance of constant vigilance to protect front-line antibiotics from abuse in settings other than hospitals.

Synercid: the first of the human-use streptogramins

An antibiotic with the trade name Synercid, which is a combination of two compounds, quinupristin and dalfopristin, entered the human clinical drug market in 1999 as an intravenously administered drug for problematic nosocomial infections. Quinupristin and dalfopristin are streptogramins produced by certain *Streptomyces* species. At the time, streptogramins were not new antibiotics, but they were new to human use. A streptogramin called virginiamycin had been used for years as a growth promoter in animal husbandry. Synercid's claim to fame is that it is effective against many of the same Gram-positive bacteria that were once susceptible to vancomycin but are no longer. When resistance to vancomycin began to emerge, especially among *Staphylococcus* and *Streptococcus* species, the need for antibiotics that would be effective against these resistant Gram-positive bacteria became particularly urgent. Synercid was supposed to be an answer to this problem.

In practice, Synercid has been something of a disappointment. First, Synercid is not effective against all Gram-positive bacteria. Notably, vancomycin-resistant *Enterococcus faecium* (VRE), a cause of postsurgical infections in immunocompromised patients, varies in its susceptibility to Synercid. Second, a side effect in some people is arthritis and general muscle pain, which can be severe. Irritation at the site of entry of the catheter also occurs in patients receiving intravenous administration via catheter. Nonetheless, Synercid has been useful in treating infections caused by vancomycin-resistant strains of *S. aureus* and *Staphylococcus epidermidis*.

Streptogramins, like the macrolides and lincosamides, act by binding to an RNA component of the large 50S subunit of the ribosome. Quinupristin and dalfopristin act synergistically. While each of the antibiotics alone is bacteriostatic, together they are bactericidal. When dalfopristin binds to the 50S subunit, it changes the shape of the ribosome, which in turn enhances the binding of quinupristin to a nearby site by about 100-fold. This binding prevents elongation of the protein chain and causes incomplete (and inactive) proteins to be released from the ribosome. The fact that their binding site overlaps that of macrolides and lincosamides has an ominous implication for development of resistance to them. As mentioned in chapter 6, a modification of the RNA in the large subunit of the ribosome can lead in a single step to resistance to all three classes of MLS (macrolide, lincosamide, and streptogramin) compounds. Thus, the useful life of Synercid could be rather short simply due to cross-selection of resistances.

Synercid also illustrates a frustrating problem that confronts scientists and farmers who use antibiotics to improve animal husbandry. In contrast to the animal-use antibiotic tylosin, which was known to be in the same macrolide family as erythromycin and thus had the potential to select for cross-resistance to human-use antibiotics, the streptogramins used in agriculture had no human-use analog until Synercid entered the market. Changes in the availability of new human-use antibiotics can place older agricultural-use antibiotics, such as virginiamycin, in jeopardy of being banned from future agricultural use because they may cross-select for resistance to antibiotics recently added to the list of human-use pharmaceuticals. European countries and others have already placed bans on virginiamycin. Studies have found animal strains of Gram-positive bacteria that are resistant to Synercid, raising questions about how long farmers are going to be allowed to use virginiamycin in the United States. The list of agricultural antibiotics that are considered safe in the sense that they do not cross-select for resistance that could impair important human-use antibiotics continues to contract.

DAPTOMYCIN—REFORMULATION YIELDS A NEW DRUG OF LAST RESORT

During the drug discovery and development process, pharmaceutical companies conduct initial screens to determine whether the lead compound is a suitable drug candidate by testing how effective it is as an antibiotic against different kinds of bacteria. Once they have a few hits, the candidates undergo formulation for dosing and stability and then testing in preclinical animal studies for safety and efficacy. The top drug candidates are then tested for human safety and efficacy in clinical trials in humans. For some drugs, preclinical testing and early clinical trials might determine that while the drug is effective, there are certain undesirable side effects. Sometimes, it might not be until later human clinical trials that adverse side effects or reactions are encountered. Once that happens, many of these potential drug candidates get set aside or even dropped by the large pharmaceutical companies that first develop them. But some small biotechnology companies may acquire the rights to the drug candidates and revisit the use of these compounds by making various adjustments in the formulation, dosing, or treatment times that might correct the problems observed in the initial clinical trials. Daptomycin is one such example of a reformulated antibiotic.

Daptomycin, like vancomycin, is a large, bulky cyclic lipopeptide antibiotic produced by *Streptomyces roseosporus*. It acts on Gram-positive bacteria by inserting into the cytoplasmic membrane and forming holes that leak ions and cause cell death. Daptomycin was originally discovered and developed by researchers at Eli Lilly in the 1980s. However, while it showed promise in early clinical trials, it later exhibited adverse side effects on skeletal muscles, including muscle pain, and subsequently was abandoned by the company. A small startup company, Cubist Pharmaceuticals, acquired the rights to the drug candidate in 1997 and worked on changing the original high-dose administration regimen to reduce some of the side effects. Cubist reintroduced daptomycin on the market under the brand name Cubicin in 2003, with a different treatment regimen that reduced toxicity to acceptable levels. In 2015, Cubist became a subsidiary of Merck & Co. Daptomycin is now considered a clinically important drug of last resort against multidrug-resistant (MDR) infections caused by methicillin-resistant *Staphylococcus aureus* (MRSA), VRE, *Streptococcus* species, and other Gram-positive pathogens.

NEW CLASSES OF ANTIBIOTICS—A FEW SUCCESS STORIES TRICKLE THROUGH

While the overall output has steadily diminished over the past few decades, massive antibiotic discovery efforts by academics and pharmaceutical companies using modern technologies have yielded a few notable successes. Most of the new antibiotics introduced are members of known classes of antibiotics, usually with modified structures that are derivatives of the parent antibiotic. These derivatives have been made possible through modern synthetic chemistry coupled with high-throughput screening technologies enabled by automated robotics capabilities. However, there have also been a few completely new classes of antibiotics discovered, mostly through rational design methods that take advantage of advanced computational modeling of compounds with their potential targets, followed by synthetic chemistry and biotechnology methods, and again high-throughput screening technologies.

Large-scale, high-throughput screening efforts for natural product leads

The way most natural products are screened for their potential as antibiotics consists mainly of growing soil microbes on a range of agar media and then looking for the ability of the colonies to antagonize the growth of selected target bacteria. Those microbes displaying antibiotic activity would then be processed to determine what component or components that they produced led to the observed antibiotic activity. The next step would be to separate the antibiotic component from the other cellular components of the microbe, and then purify it, determine its structure, and characterize its mode of action. One of the key goals of these studies was to distinguish the active component from other already-known antibiotics, in particular to determine whether the new compound was more potent or unique in its properties.

In most cases, converting a natural product produced by an environmental microbe that displays some antibiotic activity into a compound suitable for drug development can be challenging, for various reasons. The microbial source often produces the compound under specific conditions that are not always reproducible in the laboratory. The amount of compound produced can also be low or variable, and extracts containing the compound can be complex mixtures of many compounds that need to be separated through multiple purification steps using different chromatography methods. The resulting "lead" compounds usually have only weak activity and almost always have a few undesirable traits, such as instability to certain conditions like heat or acid, difficulty with penetrating into bacterial cells, susceptibility to antibiotic resistance mechanisms, and toxicity to the host. To overcome these undesirable properties, scientists carry out a series of chemical modifications altering the lead compound's structure in a systematic way. These modifications yield many new compounds, each with slight variation from the original structure. Each of these new compounds must then be screened for antibiotic activity to find those with improved activity. Hits with more potent antibiotic activity must then be screened further for the absence of undesirable traits. This process is then repeated several times to eventually yield a few semisynthetic "leads" that can be advanced to the next stage of drug development.

Because natural products are structurally complex, traditional semisynthetic methods of optimizing the original natural product often relied on a combination of biological production by the source microbe and feeding the altered chemical building blocks to the microbial cultures with the hope that the modified versions of the building blocks would be incorporated into the product. This strategy, for example, has worked very well for changing the structures of the beta-lactam antibiotics, generating may different alternative chemical features surrounding the core beta-lactam ring. Modern advances in organic chemical synthesis have now also made it feasible to synthesize semisynthetic antibiotics directly from the natural product.

As you can imagine, the whole process described above is quite labor-intensive and time-consuming. Indeed, it would not be possible or profitable for pharmaceutical companies to perform this on a large scale without modern automated robotics and high-end instrumentation platforms that enable high-throughput activity screening, chemical processing, and safety testing of each of the compounds obtained. We provide here some successful examples that have emerged from these enormous antibiotic discovery efforts.

Azithromycin—one of the triumphs of semisynthetic screening. As already mentioned in chapter 5, azithromycin is a next-generation macrolide related to erythromycin that has demonstrated considerable success as an antibiotic for treating sexually transmitted diseases. Azithromycin came out of a large semisynthetic screening effort to improve the properties of existing macrolides like erythromycin. Improving its acid stability enabled the antibiotic to be administered orally. Additional improvements included broadening its spectrum of activity for more types of bacteria; increasing its cellular absorption properties so that it is easily absorbed by the intestine and easily crosses cell membranes; and adding its extended-release formulation. Because it is more stable, is released more slowly, and lasts longer inside the body, treatment only requires a single dose, which as mentioned before is a major bonus for clinics treating sexually transmitted diseases like chlamydia and gonorrhea.

Arylomycins—natural product cyclic lipopeptide antibiotics with new bacterial targets. Screening efforts identified a new cyclic lipopeptide antibiotic, arylomycin, produced by *Streptomyces* species that acts against Gram-positive bacteria. Like vancomycin, it is too large to get through Gram-negative outer membrane porins. Since its discovery in 2002, several improved arylomycin analogs have been developed that converted the original natural product with weak antibiotic activity and limited spectrum against only a few Gram-positive bacteria like MRSA to drugs with potent, broad-spectrum antibiotic activity, including against Gram-negative bacteria. Some are especially effective against ESKAPE pathogens.

Arylomycins bind and inhibit a membrane-embedded signal peptidase (SPase) that cleaves the N-terminal signal peptide from secreted proteins. The signal peptide is a tag at the beginning (N terminus) of secreted proteins that helps to bring them to the cytoplasmic membrane so they can be transported across the membrane to the outside of the cell. After cleavage by SPase, the signal peptide remains stuck in the membrane, but the secreted protein is released into the medium. SPase is a promising target for antibiotics because it is essential for bacterial viability and growth. Of course, a worry that immediately arises with this new bacterial target is the ease with which selective pressure could lead to mutations conferring antibiotic resistance.

Everninomicins—natural product oligosaccharide antibiotics. Everninomicins are a different class of decorated oligosaccharide-derived antibiotics that were discovered to be produced by the soil bacterium *Micromonospora carbonacea.* These antibiotics inhibit bacterial protein synthesis by targeting the 50S ribosomal subunit at a site that blocks amino acid-tRNA binding (see Fig. 5.5).

What is attractive about these antibiotics, which has garnered renewed interest in them, is that they block bacterial protein synthesis by binding at a site that is distinct from where other ribosome-binding antibiotics bind. This unique binding site means that they are less likely to display cross-resistance with other ribosome-binding antibiotics. The biosynthetic genes that produce these antibiotics are complex, but medicinal chemists have begun to manipulate the biosynthetic pathways to produce various derivatives that have altered antibacterial activities and more favorable drug properties. Some are highly effective against MDR strains of Gram-positive bacteria such as MRSA.

Bedaquiline—first new class of anti-TB drugs in more than 40 years. Bedaquiline, a member of the diarylquinoline class of drugs, represents the first new class of anti-tuberculosis (TB) drugs to successfully pass human clinical trials. It gained FDA approval as part of a combination therapy for MDR TB in 2012. The lead compound was identified from whole-cell bacterial screening for inhibition of bacterial growth by a large library of chemicals, where the lead compound inhibited only *Mycobacterium tuberculosis* and no other bacteria.

Unlike other quinolone antibiotics that act on DNA gyrase, bedaquiline uniquely and specifically inhibits the growth of *M. tuberculosis* by blocking the action of the proton pump of ATP synthase, which is a critical enzyme required for ATP synthesis in *Mycobacterium*. Because ATP production is necessary for energy-requiring processes in the cell, after a few days the bacteria end up dying from the lack of ATP. What is exciting about this drug is that it is also effective against dormant (nonreplicating) *Mycobacterium* because the lower ATP stores in this metabolic state make the bacteria more vulnerable to further depletion of ATP. This property allows the treatment duration for TB to be shortened because the antibiotic clears the bacteria more rapidly than other anti-TB drugs. An added advantage to this shorter treatment duration is the reduced likelihood that resistance will arise.

MODERN COMBINATORIAL CHEMISTRY AND COMPUTER-AIDED RATIONAL DESIGN APPROACHES JOIN THE HUNT FOR NEW DRUGS

Some antibiotics, like fluoroquinolones (ciprofloxacin), sulfonamides, and trimethoprim, are completely synthetic, in that they were discovered through relatively small-scale laboratory screening of collections of starting compounds that were obtained by organic chemical synthesis. As large pharmaceutical companies ramped up their antibiotic discovery programs, they moved from directly developing the natural products to applying semisynthetic approaches based on feeding the microbes synthetic chemical precursors.

With the rising need to find new classes of antibiotics to combat antibiotic resistance, many companies realized that the structures of the compounds making up the chemical collections to be screened needed to be more diverse. So they also expanded their antibiotic discovery programs to include *de novo* (completely new) or entirely synthetic structure-based approaches to derive drug candidates with totally new chemical structures. One approach to generate large collections of compounds with diverse structures is to use combinatorial chemistry. Starting with an initial structure, a large variety of chemical groups can be added to a chemical reaction, which will create a mixture of many new compounds. Each of the mixtures is screened for antibacterial activity and other desirable properties, and if any of the mixtures shows a positive response, then the individual components in that mixture can be isolated and tested to determine which component had the desired property.

The biggest advantage of combinatorial approaches, whether using semisynthetic chemistry coupled with biological cultures or strictly organic chemistry methods, is that the number of mixtures screened can be ramped up to simultaneously test hundreds of compounds at a time. As mentioned above, the advent of high-throughput screening platforms using automated robotics and instrumentation has enabled this massive undertaking.

Linezolid and other oxazolidinones—completely synthetic antibiotics

Linezolid, first approved for commercial use in 2000, belongs to a completely new class of antibiotics called oxazolidinones. It is completely synthetic (does not occur in nature at all), and its structure is quite different from other naturally produced antibiotics since it was not based on the structure of an existing natural product. This class of antibiotics also made news in the scientific journals because they act at an earlier stage of protein synthesis than any previously known antibiotic. Like streptomycin, they inhibit the formation of the intact ribosome (the initiation complex) rather than later steps of protein synthesis. The oxazolidinones are used primarily to treat infections caused by aerobic Gram-positive bacteria that are resistant to other antibiotics, including *Streptococcus pneumoniae*, VRE, and MRSA. Linezolid is also proving to be relatively human-friendly. Even in severely ill patients in hospital intensive care wards, who often have numerous other underlying conditions, the incidence of severe side effects has been gratifyingly small.

The fact that linezolid has a different ribosomal target site from previously known antibiotics is important because these are not likely to be targets for the development of cross-resistance selected by other classes of antibiotics binding to the same site on the ribosome. Not surprisingly, however, bacterial resistance to these new antibiotics has emerged in some strains, usually due to mutation in the ribosomal RNA (rRNA) that prevents binding of linezolid to the ribosome.

Iboxamycin—structure-guided design leads to a new synthetic antibiotic

Most of the antibiotic discovery approaches discussed so far have involved mostly random exploration of different structural modifications on the starting antibiotic compound to generate a panel of potential drug candidates that can be tested for further development. An alternative antibiotic discovery approach that is much more targeted starts with the molecular structure of an existing compound that has the desired antibiotic activity but lacks other desirable properties that make it a good drug candidate, such as acid and heat stability, membrane-traversing ability, solubility, and a structure that will not be recognized by bacterial degrading enzymes. This structure-guided approach is called rational drug design because the scientist uses computer modeling simulations to help design the chemical starting materials and reactions needed to achieve a product that structurally resembles the antibiotic but is different chemically.

One such success story that has recently come from this approach is the antibiotic iboxamycin. In this case, the lincosamide antibiotic clindamycin, which targets bacterial ribosomes to shut down protein synthesis (see Fig. 5.5), was used as the starting antibiotic. The researchers were hoping to combat clindamycin resistance that occurs through drug modification, rRNA modification, target protection, or efflux pumps typically found in MDR pathogens (discussed in chapter 6). They used the structure of clindamycin bound to its bacterial target rRNA to design many new compound structures using structure-guided computer simulations, each synthesized using a different set of starting compounds and made using different chemical reactions. The resulting collection of more than 500 drug candidates was then subjected to high-throughput screening assays to yield 3 compounds with high potency, broad-spectrum activity, and efficacy in overcoming resistance against MDR bacteria, including the ESKAPE pathogens.

THE PROFITABILITY CALCULATION: AN INESCAPABLE FEATURE OF MODERN MARKETS

So, with the ruefully low number of new antibiotics advancing through the pipeline and the alarmingly rapid rise of MDR threats, why are so many large pharmaceutical companies, the implicit champions of the human side of the human-microbe race, tiring of the antibiotic discovery race and dropping out? Critics of the pharmaceutical industry have attributed this downsizing to the greed of the big industry players who would rather peddle high-profit drugs like Prozac (fluoxetine) for depression, Lipitor (atorvastatin) for heart disease, metformin for diabetes, and Viagra (sildenafil) for erectile dysfunction, which are long-term medications bringing in lots of money over the lifetime of a patient, than focus on antibiotics that may save a patient's life but are used for only a short period of time. While the profitability differential is clearly a significant factor, the decision of pharmaceutical companies to exit the antibiotic discovery arena has been based on compelling economic pressures that need to be better understood if they are to be countered.

Clearly, the relatively unattractive profitability of antibiotics has played a major role in the decision of pharmaceutical companies to pull out of the antibiotic discovery field. A critical factor in the cost of bringing an antibiotic to market is the cost of clinical trials necessary to demonstrate safety and efficacy. Time is also a problem in bringing a new antibiotic to market. Between lead development, preclinical and clinical trials, and the subsequent approval process, it can take 15 to 20 years for an antibiotic to go from the discovery phase to the market. This is not simple red tape, but is the amount of time needed for a rigorous assessment of the efficacy and safety of antibiotics that covers different age groups and people with different genetic backgrounds. Everyone agrees that such testing is necessary, but the bottom line is that it now costs more than $1.5 billion (U.S.) to take a completely new antibiotic through the testing and approval process and bring it to market (Fig. 8.1). To make matters worse, the average annual sales for recently approved antibiotics are less than $50 million (U.S.), making it nearly impossible for a company to recover the bulk of the investment in any reasonable amount of time, much less make a profit.

It is true that the profitability of antibiotics is considerably lower than that of drugs for treating neurological diseases, heart disease, diabetes, cancer, and depression. An indication of this is provided by a number that pharmaceutical companies use to estimate the likely profitability of a drug. This number is called the net present value (NPV) and is a way of calculating the return on investment (ROI) for a drug product. As in many other areas of life, a large NPV is good. In 2011, a report from the British Office of Health Economics estimated the average NPV for an oral antibiotic to be $50 million (U.S.) (up to $100 million for an injectable drug), whereas musculoskeletal medications and neurological drugs have NPVs of $1.15 billion and $720 million, respectively. The primary reason for such a comparatively low NPV for antibiotics is the low profit margin after it has been marketed, largely from low sale volumes, short prescription durations (weeklong versus yearlong treatment regimens), low prices caused by intense market competition, and restrictions imposed by infection control officers to limit antibiotic use.

Hidden behind these calculations, however, is a question that still needs to be confronted by the public and the medical community. With the current state of low ROI, how are drug discovery and development to be financed? Right now, people in developed countries, especially the United States, are paying these costs. Should

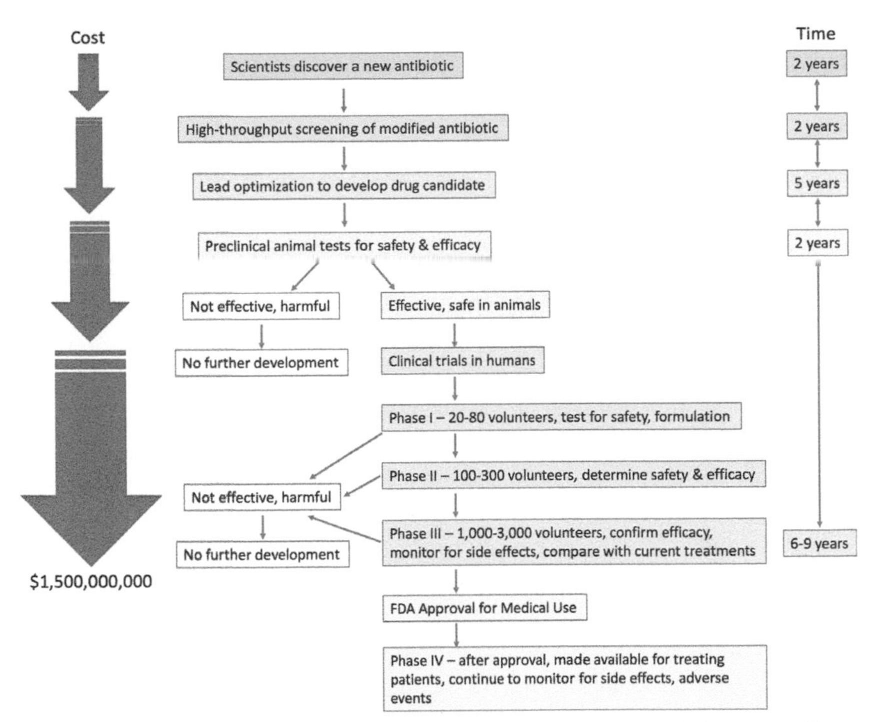

FIGURE 8.1 *The process required for approval of a new antibiotic. Shown are the key steps in the development of an antibiotic from lead discovery to animal safety and efficacy testing and then the long process of clinical trials before obtaining FDA approval for medical use. On the left is the scale of costs associated with each step and on the right is the time needed for completing each. The most challenging, time-intensive, and risky steps, where the highest attrition occurs, are during lead optimization and animal safety testing, and again at the end of phase 3 clinical trials, when filing for FDA approval of the drug for commercialization.*

these costs be spread out over a larger base and not just folded into the cost of a drug that is still under patent protection? Could the costs of developing new antibiotics be reduced without endangering public health? As a nation and a world, we have yet to make a serious attempt to confront these critical questions. A point worth making is that the pharmaceutical companies have a disingenuous argument of their own. To hear them tell it, they bear all the costs of research and development. In fact, in virtually all cases the initial discoveries and preclinical tests are made in academic or government laboratories or in small biotech startup companies. The National Institutes of Health (NIH) has grant programs for academics and small businesses to receive funding for studies through phase 1 and 2 clinical trials, but after that the cost and scale are often too high to continue. The big pharmaceutical companies are then expected to acquire these discoveries and develop them further, with the largest costs coming from the large clinical trials. Also, a large part of what is called "development" by some drug companies is not just the clinical trials, but also advertising the drug.

Will something replace these big companies that pioneered the antibiotic revolution? Will they be forced back into the race? Will we find another way to provide incentive for finding new antibiotics? The answers are still not clear, but most of us would agree that somehow the antibiotic miracle needs to be sustained.

DO WE REALLY NEED NEW ANTIBIOTICS?

The declining profitability of antibiotics has made them less and less desirable to pharmaceutical companies. Given the expense of developing a new antibiotic, some in the pharmaceutical and medical community have questioned whether we really need more new antibiotics. This startling view stems from the fact that at present, the vast majority of patients who need to be treated with antibiotics have an infection that can still be controlled with some existing antibiotic or easily made derivative thereof. The outcome of their treatment will be satisfactory, at least for the present.

The counterargument looks to the future rather than to the present. Vancomycin was hailed as the solution to MRSA infections, but vancomycin-resistant strains of *S. aureus* have already emerged. These strains have so far been susceptible to at least one other antibiotic, but the fact that *S. aureus* can become resistant to vancomycin raises the specter of the future evolution of a strain that is resistant to all available antibiotics. Similarly, the fact that *S. pneumoniae* remained susceptible to penicillin for so long was used to argue that there was no need to consider using other antibiotics for therapy. The dramatic rise of penicillin-resistant strains of *S. pneumoniae* and the concomitant rise of strains resistant to macrolides and tetracycline point out the need to expect bacteria to become resistant to a trusted antibiotic at some point in the future.

As noted in chapter 3, there are already some strains of bacteria that are extensively drug resistant (XDR); some of these are also called "pan-resistant" because they are resistant to all commonly used antibiotics. Some examples are *Pseudomonas aeruginosa*, a cause of postsurgical infections and a common cause of lung infections in cystic fibrosis patients, and *Acinetobacter baumannii*, a not-well-publicized species that is making the rounds in intensive care wards. In some cases, physicians have felt constrained to resort to antibiotics that were previously considered too toxic for general use, an ominous portent.

Perhaps a useful way to look at this problem is to ask what would have happened if back in the 1960s the medical community had decided that the first generation of penicillins and tetracyclines was adequate and stopped discovering new antibiotics. If that decision had been made, we would be in serious trouble today. True, some bacteria are still susceptible to good old penicillin and tetracycline, but resistance to these antibiotics has become so pervasive that we would now be facing a treatment crisis if the decision to stop with the first-generation antibiotics had been made.

Keep in mind that infectious diseases are one of the leading causes of death worldwide. In the developed world, infectious diseases have been demoted in the rankings over the past few decades, especially given all the advances we have made in prevention of these diseases. However, infectious diseases could come roaring back in the near future if care is not taken now. The old science fiction movies used to end with an admonition to "watch the sky." Today, in the antibiotic resistance environment we are experiencing, the admonition is "watch the hospitals."

Becoming complacent about the need for new antibiotics is a dangerous strategy. It assumes that disease-causing bacteria will remain pretty much the way they are now for the foreseeable future. Given that bacteria have surprised us in the past with their adaptability and given the strong selection pressures imposed by the widespread use of antibiotics, not only in human medicine but also in agriculture, it is unlikely that bacteria will stop evolving new resistance strategies. After all, bacteria have been here for more than 3 billion years, whereas we humans number our history in the millions of years. It is most unwise to underestimate the capabilities of such adaptable organisms.

WHAT IS TO BE DONE?

If the flow of new antibiotics from the research laboratory to the market is being impeded by economic pressures on for-profit companies, who will take up the slack? Currently, the word on the antibiotic street is that the government, in the form of the NIH and the Centers for Disease Control and Prevention (CDC), may begin to play a bigger role in the antibiotic discovery effort. In fact, the NIH, in a reorganization of its infectious diseases research funding, as part of the National Institute of Allergy and Infectious Diseases, created a new panel to consider grant proposals on antibiotic discovery and resistance. It appears that some of these moneys are not just for academic research endeavors, but also small business applications targeting biotechnology startups. Both sectors are now the primary drivers of antibiotic discovery, although most of the small companies are struggling to meet the costs of new lead discovery and clinical trials.

How can large pharmaceutical companies be drawn back into the antibiotic discovery area? One thing that needs to be acknowledged and dealt with appropriately is the fact that the pharmaceutical companies are for-profit companies with shareholders who are very interested in stock prices and profitability. One suggestion for encouraging pharmaceutical companies to reenter the antibiotic discovery area is to provide market entry rewards and other incentives for staying in the arena. Of course, this would entail increased government funding on a much larger scale than ever before (see Fig. 8.1), which in turn would be felt by taxpayers. Whatever you think about this type of proposal, it is important to keep in mind that solving the antibiotic discovery problem is not going to be simple and is not going to be cheap. What the public is willing to accept in the end comes down to how valuable the average person thinks antibiotics are for our well-being. Unfortunately, the recent massive viral pandemic and the concomitant shift in public focus toward combatting viral diseases have forced, at least temporarily, antibiotics and bacterial diseases to take a back seat to the development of viral vaccines and antivirals.

POINTS TO PONDER

In the pre-antibiotic era, people had not experienced the cures made possible by antibiotics, so their expectations were not very high. In a post-antibiotic world, people would know what they had lost, and chances are that they would not be happy about it, to put it mildly. The loss of an effective cure would be a historical first, and no one has any idea how the average person would respond to such a change in fortune. What do you think the public's reaction would be to a situation where cures that had been taken for granted were suddenly no longer available?

How do you suggest we go about ranking our public priorities? Examples are the need to balance spending money to aggressively pursue the discovery and testing of new antibiotics with other claims on the public purse, such as improving education, building roads, and so on. Like it or not, such choices are going to be made and are already being made, but so far, the public has shown very little interest in this problem.

There is more to the antibiotic discovery problem than profitability. There is also the problem of the blasé attitude of the medical community toward bacterial infections. If most physicians do not take bacterial diseases seriously, why should anyone else? The complacent attitude of many physicians had its origin, ironically, in the initial elation over the success of antibiotics and the fact that humans seemed to have gained, for the first time in history, the upper hand over our microscopic enemies. During the 1960s, physicians and patients alike had every reason to be complacent about the continued efficacy of antibiotics. What would the loss of antibiotic cures do to public confidence in the medical profession?

9 The Battle Continues To Combat Antibiotic Resistance: Alternative Approaches

Despite our best efforts, there is a very real possibility that combatting antibiotic resistance is an arms race that we cannot win. The current rate of antibiotic development simply might not be able to keep pace with the emergence and spread of antibiotic resistance. Fortunately, not all hope is lost. When the game becomes unwinnable, we can just change the rules! All we need to do is to change our current approach toward drug discovery and to develop alternative treatment strategies. In this chapter, we will discuss some of the advances in these areas, including how insights from genome sequencing and bioinformatic analysis are enhancing drug discovery; how the development of new treatment and prevention modalities might decrease the reliance on traditional small-molecule antibiotics; and how returning to age-old avoidance practices like good hygiene, exposure reduction, and vaccination might reduce the need for antibiotics in the first place.

NEW APPROACHES TO DRUG DISCOVERY

Since the early 2000s, advances in genome sequencing technology have enabled scientists to determine the genome sequences of thousands of different bacterial strains. Bioinformatic analysis of these sequences has opened the door to new ways of combatting disease-causing pathogens. For example, analysis of the genomes of natural product-producing microbes has expanded the breadth of natural products that can be screened for drug discovery. Accompanying these new genome sequences

Revenge of the Microbes: How Bacterial Resistance Is Undermining the Antibiotic Miracle, Second Edition.
Brenda A. Wilson and Brian T. Ho.

has been a new wave of microbial pathogenesis research focused on understanding the molecular basis of virulence and virulence regulation. Through comparison of the gene composition of closely related virulent and nonvirulent bacterial strains, key genes required for pathogenesis can be identified. These key virulence-related gene products can then be used as targets for small-molecule inhibitor screening. Although there have been only a few notable successes so far, hopefully they are indicative of the potential of this approach.

Genome sequencing and mining microbial genomes for drug discovery

One of the exciting discoveries emerging from the bioinformatic studies of the thousands of sequenced bacterial genomes is the realization that many of the known antibiotic-producing microbes contain numerous additional genes encoding still-to-be-characterized natural product biosynthetic pathways. Indeed, known antibiotic-producing bacteria like *Streptomyces* and *Bacillus* species and antibiotic-producing fungi like *Penicillium* have the capacity, at least gene-wise, to produce more than 10-fold the number of potential natural products than previously predicted. This finding prompted scientists to explore the possibility of "mining" these microbial genomes for new natural products that can then be screened for antibacterial activity.

Microbial genome mining hinges on the fact that most if not all the biosynthetic genes responsible for producing a particular natural product are usually located near each other in the genome. This grouping of genes is referred to as a biosynthetic gene cluster. The genes encoding enzymes that produce the antibiotic's core structure are usually highly conserved, while genes encoding enzymes that chemically modify the core structure are more variable and are typically responsible for the structural diversity of the final product. Researchers first identify novel antibiotic biosynthesis gene clusters by searching for the groupings of the conserved genes. They can then analyze the associated genes to computationally predict natural product structures.

One of the biggest early challenges encountered when exploiting these novel biosynthetic gene clusters is that microbes do not always express these genes under standard laboratory conditions. It is therefore necessary to make additional genetic modifications to place these genes under inducible promoters or even to move the entire gene cluster into a more manageable organism. Modern molecular biology technologies have largely streamlined these processes, making it easy for researchers to obtain the high yields of the natural products needed to screen for antibiotic activity.

Lifestyles of the small and vicious—how understanding microbial physiology helps to identify new targets

Recently, antibiotic discovery has moved in a new direction, which has generated a great deal of excitement in the field. Instead of targeting bacterial pathways that are essential for a bacterium to survive, it may be possible to instead focus on the pathways required for colonizing or infecting the host. This new drug target paradigm is enabled by comparative genome analysis tools that can identify sets of genes that are expressed only when the bacteria are under "infection" conditions. Examples of such new targets include biosynthetic genes responsible for formation of capsule or biofilm, regulatory pathways that control toxin production and secretion, and genes conferring cellular adhesion properties. Additionally, these virulence pathways intersect not only with each other but also with bacterial antibiotic resistance mechanisms. Therefore, by screening for inhibitors of virulence pathways, drugs that increase pathogen susceptibility to existing antibiotics can be found, along with direct inhibitors of virulence.

One important regulatory system employed by many bacterial pathogens is quorum sensing. Simply put, quorum sensing is a cell-to-cell communication system that enables bacteria to sense the number of similar bacteria in their vicinity. Once a certain cell density (i.e., a quorum) is reached, the bacteria turn on a select group of virulence or metabolic genes. One of the notable success stories arising from screening for compounds targeting quorum sensing is a natural product from witch hazel called hamamelitannin. Originally identified computationally as a putative inhibitor of a quorum-sensing regulator controlling biofilm formation in methicillin-resistant *Staphylococcus aureus* (MRSA), hamamelitannin was found to make MRSA biofilms more susceptible to vancomycin treatment.

Another success story of directly targeting virulence is the synthetic small molecule virstatin. This compound was originally identified through screens for inhibitors of cholera toxin production in *Vibrio cholerae*. It was subsequently found to bind and inhibit the regulatory protein ToxT, a direct transcriptional activator of the genes encoding cholera toxin, and was also shown to protect mice from diarrhea and intestinal colonization by *V. cholerae*.

Ironically, the biggest drawback of these highly targeted strategies is that they are so highly targeted. As a result of specific targeting of virulence mechanisms in a particular strain, other strains of the same bacteria that have evolved slightly different regulatory systems may remain completely unaffected. For example, shortly after the discovery of virstatin was first reported, toxigenic strains of *V. cholerae* were discovered that carried a variant ToxT gene that was unaffected by virstatin. Additionally, the high specificity of these drugs means that they are not a suitable replacement for antibiotics as we currently use them. Right now, first-line antibiotics often need to be administered before the specific pathogen has even been identified, so these specific virulence inhibitors end up being relegated to very niche situations. But in a world where the typical first-line antibiotics do not exist anymore because of antibiotic resistance, those situations might not be so niche.

NEW ALTERNATIVE STRATEGIES ON THE HORIZON

In addition to exploring new drug development strategies, researchers are also exploring entirely new treatment strategies that do not rely upon antibiotics. We discuss here a few interesting examples of these new research directions.

Probiotics and fecal microbiota transplantation

Unfortunately for some individuals, perturbations of the gut microbiome through antibiotic treatment can lead to life-threatening illness. Broad-spectrum antibiotics can result in the loss of protective "good" bacteria, which in turn can lead to overgrowth of "bad" bacteria that can cause damage to the host. Dental caries, bacterial vaginosis, and inflammatory bowel diseases, like Crohn's disease and ulcerative colitis, are all examples of diseases likely resulting from disruption of the normal microbial community composition. One proposed treatment strategy for these diseases that is slowly gaining traction is to add back enough of the "good" bacteria to restore the healthy state of the microbiome that naturally suppresses the growth of the "bad" bacteria.

One area where microbiome replacement appears to be having success is in the case of recurrent colitis (diarrhea, abdominal cramping, inflammatory bowel) caused by overgrowth of *Clostridioides difficile* (formerly known as *Clostridium difficile*). This condition is a common complication of antibiotic therapies that kill off significant portions of the normal gut microbiota. The condition is particularly problematic because after treatment of the *C. difficile* infection with a course of antibiotics, roughly 30% of patients suffer from a recurrence of the infection. If recurrence happens again after another round of antibiotics, the patient becomes a prime candidate for fecal microbiota transplantation (FMT) therapy, where a fecal sample from a healthy donor is given to the patient, usually either administered orally as capsules containing fecal material (poop pill) or delivered rectally via enema or colonoscopy. Reported success rates are currently as high as 90% for treatment of these recurring *C. difficile* infections.

So far, FMT has been successfully applied only in situations where a damaged microbial community is being repaired. However, there is hope that it might be possible to take this approach one step further. With better understanding of the intricate intercellular interactions both among microbes and between the microbes and the host, it might be possible to supplement our microbiomes with additional "good" bacteria that proactively prevent infection by pathogens (also known as probiotics). Alternatively, it may be sufficient to supply our microbiomes with nutrients or conditions that promote the growth of the "good" bacteria (also known as prebiotics). Research is still ongoing to determine the extent of the health benefits afforded by currently available probiotics and prebiotics, but, at the minimum, side effects appear to be uncommon, with most adults able to safely include prebiotic foods and probiotic or dietary supplements in their diets.

Antiadhesion therapy

One of the critical steps required for infection is the initial colonization of the host, where bacterial cells adhere to cellular or mucosal surfaces inside the body. Consequently, blocking this binding from occurring will prevent colonization and minimize infection. The host immune system uses this feature as one of its primary strategies for keeping the body free of infection. Antibodies made by immune cells bind tightly to the surface of the bacteria and prevent them from attaching to host cells. Additionally, antibodies can also bind to bacterial products (such as secreted toxins) to prevent them from harming the host. These antibodies can be collected from blood and stored as antiserum. In the event of a dire situation, such as an untreatable multidrug-resistant (MDR) infection, this antiserum can be administered to the patient to assist their immune system in clearing the infection.

This approach is called passive immunization, and it has been particularly successful in treating toxin-mediated diseases, such as diphtheria and botulism, where the antibodies (antitoxins) bind and neutralize the toxins to prevent toxin action. It is also useful in cases where the patient is immune compromised and unable to generate their own antibodies against the MDR pathogen. The downside of passive immunization is that the antibodies used need to be humanized, meaning that they either need to come from humans or need to be made to look like antibodies from humans. Because nonhuman antibodies are sufficiently different from their human counterparts, our immune systems will recognize them as foreign entities. Consequently, they can elicit adverse immunological responses.

Instead of using antibodies, another way to block bacteria or their toxins from attaching to cells is to use a competitive inhibitor. Bacterial adhesins and the binding domains of toxins bind to specific chemical groups, typically sugars, decorating host cell surface receptors. Following administration of high concentrations of free-floating versions of these sugar groups, bacteria and their toxins will bind to them instead of the sugars on the host cell surface. This type of treatment has progressed to clinical trials for a couple different bacterial toxin-mediated diarrheal diseases. This approach can also potentially be used to deal with bacterial biofilms formed by some foodborne pathogens.

A major difficulty of using sugars to prevent bacterial adhesion is knowing what sugars to use, as bacterial adhesins bind to a wide variety of molecules. However, a perhaps bigger complication is the fact that many bacteria readily metabolize these sugars, effectively removing them from the extracellular environment. An alternative to using sugars is to use a soluble recombinant version of the host cell surface receptors. For this strategy, the part of a gene encoding the host cell surface receptor that is recognized by the bacteria is cloned and expressed as a soluble recombinant protein, purified, and then administered to an infected patient, either through injection or orally in an acid-tolerant formulation. As in the case of the free-floating sugars, the bacteria bind to the soluble receptor instead of host cells.

Bacteriophage therapy

We discussed in chapter 7 how some viruses that infect bacteria (called phages) can spread antibiotic resistance genes through horizontal gene transfer from one bacterium to another. However, one key aspect of the phage life cycle that we did not put much emphasis on is that for most phages, propagation of the phage involves the lysis (and destruction) of the host bacterium. Scientists are currently exploring ways to use these phages to control pathogenic bacterial populations.

The idea of using phage therapy to treat bacterial infections is not actually new. In fact, it was attempted as a treatment method in the early 1900s, even before antibiotics came onto the scene in the 1940s. The problem that plagued phage therapy developers at the time was that it was simply too unreliable. Because most phages have such a narrow spectrum of activity, usually acting against only certain bacterial strains, it was often difficult to predict which clinical strains would be susceptible. So, once antibiotics were discovered, development of phage therapy completely fell by the wayside. But with the rise of antibiotic resistance, this approach has been given renewed attention. Experimentally, phage therapy has been effective for topical infections with staphylococci and streptococci, as well as *Bacillus anthracis* (Fig. 9.1),

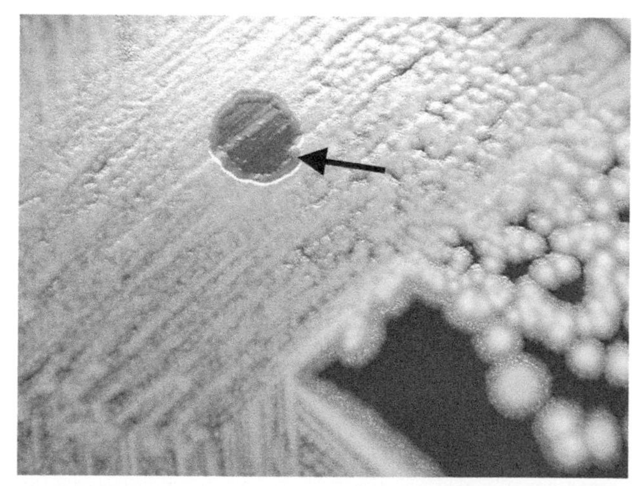

FIGURE 9.1 *Bacteriophage killing of a lawn of* Bacillus anthracis *colonies cultured on a sheep blood medium-containing agar plate. The black arrow indicates a cleared, circular area where bacteriophages lysed the bacteria. Eventually, the zone of clearing would spread through the entire bacterial lawn, due to the bacteriophages killing all of the bacteria on the plate. Courtesy of CDC-PHIL (ID# 1883/CDC/Larry Stauffer, Oregon State Public Health Laboratory).*

and in the past few years, the U.S. Food and Drug Administration (FDA) has approved a number of phase 1 and phase 2 clinical trials for phage therapy applications against *Pseudomonas aeruginosa* in cystic fibrosis patients with MDR lung infections, *S. aureus* in wound infections, *Escherichia coli* in urinary tract infections, *Acinetobacter baumannii* in MDR infections, and multiple ESKAPE pathogens in chronic prosthetic joint infections.

Bacterial parasites . . . of bacteria

Just as phages are viruses that infect bacteria, it turns out that there are a few bacteria that infect other bacteria as well. First described back in the 1960s, *Bdellovibrio bacteriovorus* is a Gram-negative, comma-shaped bacterium that actively attacks and physically invades the periplasmic space of other Gram-negative bacteria. Once inside their bacterial host cell, they permeabilize the host's inner membrane and use its cellular contents as nutrients to replicate and divide. Once the host cell has been exhausted of resources, the *Bdellovibrio* cells break out of the host cell and seek out new prey.

Preliminary studies using these predatory bacteria have shown great potential in several different animal host models; however, we are still quite far from any actual human applications. Although rapid progress is being made toward mechanistic understanding of its parasitic behavior, very little is known about what kind of interactions these predatory bacteria might have with a human host. Additionally, because these predatory bacteria are themselves living, growing bacteria, they would likely be flagged by most of our current tests for bacterial contamination in food and pharmaceuticals. Therefore, employing these bacteria as an antibiotic alternative would require a complete overhaul of our current regulatory and safety monitoring systems.

CHANGING HABITS—THE IMPORTANCE OF HYGIENE AND AVOIDANCE

Another important factor to consider is that over the past 80 years, since antibiotics first burst onto the clinical scene, we have learned a lot about preventing infections, knowledge that people in the pre-antibiotic era did not have. For example, our ability to prevent the transmission of infection in hospitals by using disinfectants and antiseptics and, in some cases, by isolating patients has improved considerably. Of course, this assumes that all nurses and physicians are fastidious about cleansing their hands between patient interactions and about properly using gloves. Unfortunately, many hospital staff members cut short some hygienic practices because of complacency or carelessness stemming from the increasingly rushed schedules that burden hospital personnel.

The introduction of alcohol-based disinfectant lotions that allow hospital staff members to disinfect their hands without having to find a sink has been an important development. Staff members can use these readily accessible, wall-mounted dispensers or simply carry a tube of lotion that they can apply while walking from one place to another. Infection prevention and control practices in the health care community and the general population as a whole have improved compared to the pre-antibiotic era, at least in more developed regions of the world. Moreover, improved nutrition and less crowded conditions have improved the robustness of the defenses against infection so that we do not get sick as readily.

Another advancement that was not available in the pre-antibiotic era and that may help to save surgery from the worst effects of antibiotic-resistant bacteria is the advent of less invasive forms of surgery, such as surgery that requires only tiny cuts and laser surgery that requires no breaching of the skin at all. These types of less invasive procedures, along with techniques that allow more rapid operations and thus reduce the opportunity for bacterial contamination of open surgical wounds, will help significantly to prevent postsurgical infections. Plastic implants that discourage bacterial growth are also an encouraging development. On the cancer front, new "smart cancer drugs" that target the tumor cells and do not decimate the immune cells that keep invading bacteria under control will lessen the risk of a cancer patient developing an overwhelming bacterial infection. Despite these numerous advances, we are still very much dependent on antibiotics to control infections in hospital settings.

THE ROLE OF ANTISEPTICS AND DISINFECTANTS

Ultimately, the best and most effective strategy for preventing infections is to stop the pathogen from entering the body right from the outset. As such, the importance of reducing the number of bacteria we are exposed to should not be discounted. After all, introduction of water treatment practices of filtration and chlorination and sewage waste disposal systems in the early 1900s greatly reduced the incidence of diarrhea by *V. cholerae*, typhoid fever by *Salmonella enterica* serovar Typhi, and dysentery (bloody diarrhea) by *Shigella* species. Likewise, handwashing, applying disinfectants or antiseptics, and sterilizing instruments are well established as critical practices in hospital settings to reduce disease occurrence and spread.

In chapter 1, we initiated a discussion about the application of antiseptics and disinfectants as highly effective antimicrobials against not just bacteria but also other microbes and viruses. A drawback that we noted is that these compounds are generally too toxic to be administered internally in humans or animals. Nevertheless, with the rise of antibiotic resistance and the increasing need for reducing exposure to MDR pathogens, external use of these antimicrobials is moving into the spotlight. Unfortunately, just as progress in identifying new antibiotics is slowing, progress toward the development of new non-antibiotic antibacterial compounds is also slowing.

New uses of antiseptics and disinfectants—triclosan

Antiseptics are already being used as an important part of the strategy for combatting MRSA strains. For decades now, the phenolic compound triclosan has been incorporated into many consumer products, such as hand soaps, toothpaste, shaving cream, cosmetics, toys, athletic clothing, food packaging, and surgical cleaning supplies, to reduce microbial contamination and presumably make those items safer. For some consumer products, like toothpaste, there is evidence for benefit in preventing gum disease (gingivitis).

At high concentrations, triclosan acts as a microbicide that disrupts multiple membrane targets and kills microbes. However, at lower concentrations typically found in commercial health care products, triclosan appears to act more as a bacteriostatic antibiotic. Unlike other antiseptics, it has a specific bacterial target, an enzyme involved in fatty acid synthesis. A structural analog of triclosan is hexachlorophene, which is an active ingredient in many deodorants that inhibits the odor-producing activities of skin bacteria.

One increasingly important application of triclosan is to impregnate it into plastic implants and indwelling catheters. Bacterial biofilms that form on medical plastic devices are becoming a very serious medical problem. Since biofilms that form on these devices are seldom directly treatable with antibiotics, the device must be removed. For a venous catheter, this is a simple procedure: pull the catheter out of the patient's body. However, in the case of plastic heart valves or other internal implants, a surgical operation is required to remove the contaminated device. Then yet another operation is needed to reimplant a new device after antibiotics have been used to eliminate any remaining bacteria at the site. The need for these additional procedures can be avoided by impregnating the plastic of these devices with triclosan to prevent biofilm formation in the first place.

An amusing story about "antibacterial" products features triclosan. During the late 1990s, when antibacterial plastic cutting boards and plastic toys suddenly appeared in stores all over the United States, consumers doubtless assumed that something new had been added to these products. But this was not the case. Triclosan had been added to plastic products for years as a retardant of bacterial activities that can make plastics brittle and thus reduce their shelf life. All that changed between the cutting board that had no health claim and the one that was "antibacterial" was the label. Some advertising genius had simply decided to take advantage of the public's anxiety about "germs" by pointing to a compound that was already in the product.

Recent health concerns about the potential detrimental effects of absorbing and accumulating triclosan through skin, mucosa, and other bodily fluids from so many sources led the FDA in 2016 to ban the use of triclosan in over-the-counter antibacterial soap products and leave-on products such as hand sanitizers. However, while medical applications of triclosan are still in effect and prevalent, the FDA issued a final rule in 2019 that companies cannot use triclosan (as well as a number of other bioactive ingredients) as an antiseptic in over-the-counter health care products without first conducting extensive premarket review. Part of this decision was based on the lack of sufficient data regarding safety and efficacy of the so-called bioactive ingredients. However, another reason cited was the concern that antiseptics like triclosan pose a risk for development of antibiotic resistance, which would undermine its utility for important medical applications.

Resistance to antiseptics and disinfectants

In an era where antibiotic resistance is a widespread concern, it is worrying to contemplate that many of these resistant bacteria are also becoming resistant to antiseptics and disinfectants. As disinfectants are increasingly being relied upon to decontaminate hospital, nursing, and daycare settings, there is a worry that we may be selecting for microbes carrying resistance to these disinfectants. While the general public seems to be awakening to the importance of protecting antibiotics by preventing their overuse, the dangers of overusing antiseptics and disinfectants are not on most people's radar.

Resistance to antiseptics and disinfectants is still poorly understood, but it certainly does occur. This may be surprising because, unlike antibiotics, antiseptics and disinfectants typically have multiple cellular targets, and it is extremely improbable if not impossible for a microbe to change all its susceptible proteins and DNA to become resistant. It is, however, possible to prevent the antiseptic or disinfectant from reaching its cellular targets. For example, some bacteria have managed to become resistant to some types of antiseptics, such as quaternary ammonium compounds (QACs), which attack bacterial membranes. In general, QACs are inherently less effective against Gram-negative bacteria than Gram-positive bacteria, likely due to the lipopolysaccharide in the outer membrane of Gram-negative bacteria preventing the hydrophobic QACs from accessing the membrane. Interestingly, a common resistance mechanism that some Gram-positive bacteria employ to deal with QACs is to express efflux pumps in their cytoplasmic membrane, presumably to pump the disinfectant out of the cell cytoplasm. It is still unclear mechanistically how this action confers resistance, considering that QACs are thought to act mainly by dissolving membranes. Whatever the explanation, these pumps are not only effective in protecting the bacteria from QACs; genes encoding these pumps have also been found on plasmids, meaning that they are already primed for horizontal gene transfer.

The discovery that resistance to antiseptics and disinfectants can develop is disturbing because disinfectants and antiseptics are a vital line of defense against microbial infections. The fact that such resistance genes can also be linked genetically to antibiotic resistance genes on transmissible elements is even more disturbing because it means that use of disinfectants and antiseptics might end up directly selecting for maintenance of genes that confer resistance to antibiotics. Resistance to both antibiotics and antiseptics could be a particularly deadly combination, not only for hospital patients but also for sick people in the community.

Fortunately, there seem to be two disinfectants to which bacteria may not be able to become resistant: alcohol and good old household bleach. Alcohol and bleach are so general in their action that it is difficult if not impossible for bacteria to become resistant to them. It may not be an accident that hypochlorous acid, the active component in bleach, is also an antibacterial compound used by human immune cells to destroy bacteria as part of the body's defense system.

THE GROWING IMPORTANCE OF VACCINES AND PASSIVE IMMUNIZATION

Vaccines are hands down the most effective and inexpensive medical intervention to prevent infectious diseases. Compared to the cost of treating a severe MDR infection that has already caused significant damage to the patient, the cost of getting vaccinated is by far a better bargain. Indeed, vaccination prevents the pathogen from establishing an infection in the first place so that minimal, if any, damage occurs. Importantly, for infections that are untreatable due to MDR, vaccination may be the only way to protect individuals from getting the disease.

Over the past couple of decades, major technological advances and our improved understanding of how the immune system works, as well as new insights into the bacterial and host factors contributing to virulence, have enabled the development of exciting new vaccines. There are now many strategies available to enhance the immune response to vaccines such that the appropriate type of immune protection is mounted for each type of pathogen, and importantly, such that long-lasting immunity is elicited. One example, the *Haemophilus influenzae* type b (Hib) vaccine, which is now recommended by the Centers for Disease Control and Prevention for all children under 5 years of age, has made a tremendous contribution to controlling childhood diseases such as otitis media (ear infection) and meningitis. Hib is a glycoconjugate vaccine made from the capsule polysaccharide of the bacterium linked to a carrier protein that helps elicit a much stronger immune response that also lasts much longer.

For another example, until recently, diseases caused by *Streptococcus pneumoniae* usually responded well to penicillin, but resistance to penicillin has increased to the point where treatment failures in earache cases are becoming more and more common. Currently, immunization with the pneumococcal vaccine is the best way to prevent severe pneumococcal disease, including pneumonia, meningitis, and sepsis. The current pneumococcal polysaccharide-based vaccines consist of purified polysaccharides from as many as 23 different serotypes of bacteria that are conjugated (attached) to an inactive mutant of diphtheria toxin as a carrier protein. One notable limitation of these vaccines is that they cannot prevent all pneumococcal infections, since they act only against a limited subset of the strains circulating in the population. They also are not as effective in children under the age of 2, due to the immune systems of young children not being fully developed yet.

We mentioned earlier in this chapter that passive immunization has been used effectively to treat the toxin-mediated diseases caused by diphtheria toxin and botulinum neurotoxin, where early administration of antisera against the toxin (antitoxin) can neutralize the toxin and block further damage to the host. Passive immunization has also been used effectively to treat untreatable cases of MDR infections, as well as cases where the patient is immunocompromised and unable to mount their own immune defense. In the future, similar passive immunization strategies may become a more prevalent solution for MDR diseases caused by pathogens for which full immunization programs are not in place.

POINTS TO PONDER

For us humans, the ability of bacteria to become resistant to antibiotics poses a serious problem. We can, and will, try to keep pace with the development of bacterial resistance by developing new antibiotics, but is it possible that we are in a race with bacteria that we will ultimately lose? Could we one day return to a pre-antibiotic era? Public concern about the declining efficacy of antibiotics, the lack of new ones on the horizon, and where this trend might lead is understandable. Even the more optimistic among us still believe that the battle between humans and bacteria is far from over. The best we can hope for is a running standoff. But that would be far better than losing antibiotics altogether. The keys to future success in saving antibiotics or getting new ones are knowledge and the willingness of the public to take an informed interest, not only in preserving the efficacy of the antibiotics we have now, but also in ensuring that future discovery and development of new antibiotics continues and that innovative alternative strategies are pursued.

10 Building a Viable Framework for Tackling the Escalating Antibiotic Resistance Crisis

In this book, we have attempted to build a comprehensive picture of the current antibiotic resistance crisis. Losing access to working antibiotics would effectively mean undoing decades of advances in modern medicine and societal development. Considering how advanced our understanding of antibiotics and their mechanisms is, it is rather appalling how few new drugs are trickling down the pharmaceutical pipeline. While it is relatively easy to identify many factors limiting antibiotic discovery and development or driving the spread of difficult-to-treat multidrug-resistant (MDR) infections, tackling and overcoming these roadblocks is not so trivial. As we pointed out in chapter 8, the current monetary return on investment for successfully bringing a new drug candidate to market is simply too low to be viable. What we need are better ways of translating the obvious societal benefits of having functional medicines into tangible returns valued by investors, whether it be by providing better financial incentives or by having the beneficiaries become direct investors themselves. In this chapter, we will consider the various barriers and stakeholders involved in dealing with the antibiotic crisis as we chart our path forward.

Revenge of the Microbes: How Bacterial Resistance Is Undermining the Antibiotic Miracle, Second Edition.
Brenda A. Wilson and Brian T. Ho.

THE STAKEHOLDERS IN COMBATTING THE ANTIBIOTIC RESISTANCE CRISIS

Antibiotic resistance is a universal problem that affects everyone, so naturally it makes sense for everyone to contribute to the cause, right? Unfortunately, it is not so simple. First of all, who exactly is *everyone*? Technically, all human beings are potential patients, so in a sense we can all feel some unified sense of urgency, but different people represent different interest groups, each with their own perspectives, motivations, and concerns. Solving the antibiotic crisis will require mutual understanding and universal buy-in of all parties.

Who should be responsible for drug discovery and development? Who should foot the bill?

Developing novel antibiotics is a costly endeavor. Traditionally, large pharmaceutical companies have taken up the mantle of investing in the development of new drugs. However, falling profit margins for antibiotics in recent decades have resulted in these companies shutting down their antibiotic discovery efforts, leaving more and more of the burden on academic and government laboratories or small biotech startups, all of which are funded primarily through federal granting agencies. Generally, the hope has been that larger companies would step in during the later stages of drug development to translate the short list of promising new drug candidates into actual medicines. Unfortunately, the limited profit potential of antibiotics has meant that many of these potential leads are not even being pursued for further development. Clearly, the current economic and business models for antibiotic development do not work.

One might then ask, "Why can't academic or government research institutes take up the mantle?" To a certain extent, they already have. However, most academic institutions are centered around education and not-for-profit basic science research. As such, their involvement is predominantly limited to identifying new drug leads or innovating alternative treatment modalities. Finalizing the development and approval process of a new drug candidate requires so much more investment. To produce drugs at scale and conduct extensive clinical trials, the investment capital and infrastructural capacity of large pharmaceutical companies are essential. Building such capabilities within academic institutions is simply not feasible, at least not anytime soon. Consequently, if we want to make real actionable change, what we need are changes to operating business models and adjustments to current economic and regulatory policies that will enable us to better utilize our existing infrastructural systems.

Providing incentives—to push or to pull?

When formulating economic incentives to encourage pharmaceutical companies to engage in antibiotic development, there have been two main approaches. The first, which has been the dominant approach for decades, involves providing financial incentives for companies to invest in antibiotics, either by lowering the barrier to entry through tax credits and helping to form public-private business partnerships or by directly funding antibiotic development initiatives through research grants or contracts. These so-called "push" incentives serve to push companies toward getting into the antibiotic development space. One example of such an incentive is a bipartisan bill introduced before the United States Congress in 2021. The Pioneering Antimicrobial Subscriptions to End Upsurging Resistance (PASTEUR) Act was an amendment to the National Defense Authorization Act. If passed, the bill would provide monetary incentives for antibiotic development by awarding drug developers contractually agreed-upon funding to discover and develop new antibiotics.

Unfortunately, many "push" incentives have a critical flaw. Although they are great for generating interest in short-term investment, they do not actually encourage sustained efforts to see drug candidates all the way through the later stages of development and testing. At the core of the issue is the fact that these financial incentives do not sufficiently cover the most expensive part of drug development, completing the clinical trials needed for regulatory approval. Indeed, more than 75% of current drug candidates are stuck in development limbo due to financial limitations. At the end of the day, someone still needs to bear this cost, and pharmaceutical companies have made the determination that the potential profit of bringing the drug to market will not cover it.

In response to this dilemma, there has been a shift toward a second incentive approach. In contrast to "push" incentives that only drive initial investment, so-called "pull" incentives are intended to sustain interest throughout the entire development process to yield a viable product. The key lies in sufficiently increasing the financial payoff for successfully bringing a new antibiotic to market to overcome the steep development costs. These "pull" incentives come in the form of monetary prizes, patent buyouts, or transferable market exclusivity extensions (i.e., extending market exclusivity of another drug that a company might have). Additionally, these rewards can be linked to sustained sales, creating further incentive for companies to maximize distribution of their final drug product. Ultimately, a combination of both push and pull incentives will likely be needed to drive the rebuilding of our antibiotic development pipelines.

Other stakeholders weigh in

Funding organizations and pharmaceutical companies are not the only parties with a say in the antibiotic resistance crisis. Several other interest groups also play an important role in determining the policies concerning antibiotic development, responsible antibiotic usage, and the spread of antibiotic resistance.

Patients and their doctors—best treatment versus antibiotic stewardship. A core part of our current strategies for maintaining the efficacy of antibiotics and limiting the spread of antibiotic resistance involves imposing restrictions on their use. The rationale for these stewardship policies is that through the limiting of microbial exposure to the drugs, resistance will not have the opportunity to be selected. However, the need to limit antibiotic use is directly at odds with the needs of afflicted patients and the doctors trying to treat them. One major issue is that frequently, by the time a patient is brought into the emergency room, there is little to no time for samples to be sent to clinical laboratories for full diagnosis or even to identify the specific pathogen so that more targeted treatment strategies might be enabled. If the patient is not treated immediately with a broad-spectrum antibiotic, there is a strong chance that the patient will die.

Generally, physicians are fully aware that increasing antibiotic resistance in pathogenic bacteria is a serious concern in these urgent cases. However, they and their patients are essentially forced into an impossible situation where they must choose between risking one patient dying right now versus the chance that hundreds or thousands of patients may become unsavable later if antibiotics stop being effective. Although this ethical dilemma likely has no true solution, efforts can be made to mitigate the downsides of either option. For example, investments in rapid diagnostic technologies and tiered intervention strategies may give doctors more time to implement targeted therapeutics. Meta-analysis of hospital data (e.g., patient demographics, hospital conditions, resource distribution) can also be implemented to develop better treatment options. Alternatively, it may be possible to screen all incoming hospital patients for carriage of certain MDR bacteria and to isolate those patients testing positive, thereby minimizing the exposure of other patients to resistant strains in hospitals.

New multifaceted measures will not come easily. Physicians and other health care workers will likely require additional training; development of new technologies and treatment strategies will take time and effort; patient screening will add inconvenience; and of course, all these measures will cost money to implement. However, with open dialogue and continued education, it is possible to implement effective changes while minimizing costs.

Hospital administrators—setting effective practices. Hospital administrators have long bemoaned the ever-rising cost of treating patients with antibiotic-resistant bacterial infections. There are several major expenses beyond the obvious elevated price of using intravenously administered antibiotics. First, the extra days spent in the hospital when infections are not brought under control in a timely manner consume a great deal of additional supplies, space, equipment, and hospital staffing. When infectious diseases become involved, costs are compounded further as patients now need to be isolated and equipment needs to be properly decontaminated before reuse. Furthermore, untreated infections can cause systemic damage to the patient's organs like the heart, lungs, or kidneys, which in turn necessitates additional treatments or even surgery. Unfortunately, insufficient insurance coverage in many cases contributes to an even greater burden for hospitals as they must absorb these uncovered costs. This burden is particularly impactful on hospitals serving underprivileged and resource-challenged communities.

Rising concern among hospital administrators over the cost consequences of antibiotic-resistant bacterial infections has had at least one good outcome for patients. A greater importance has been placed on infectious disease control specialists in hospitals. These specialists are now doctors rather than low-level technical support staff (as they were in the past). They have collectively initiated new efforts to reinforce the use of simple but effective hygienic measures, such as handwashing and wearing personal protective equipment (gloves, masks, etc.). Over the years, these obvious measures were seemingly forgotten in an era when antibiotics naturally made up for any lax hygienic practices. Infectious disease specialists have even begun, in some settings, to preemptively isolate patients with suspected antibiotic-resistant bacterial infections. Such practices were originally limited to patients with highly contagious diseases such as tuberculosis, meningitis, or viral hemorrhagic fever. That antibiotic-resistant bacteria are being treated with the same level of caution is a clear sign that hospital administrators are taking antibiotic resistance seriously.

Insurance companies—motivating forces for cost reduction. One of the many economic factors cited by pharmaceutical companies for their retreat from the antibiotic discovery and development space is the pressure placed on health care systems and pharmacists by insurance companies to reduce overall medical costs. As part of these cost-cutting measures, insurance companies are refusing to cover the costs of using name-brand drugs when less expensive generic drugs are available, a move that can cut deeply into the profit margins of the original drug developers. When the initial patent for a newly developed drug expires, other companies are free to produce and market the drug as well. These generic versions are required to meet the same quality and potency standards as the original. However, because these companies can skip the drug discovery and development process and do not need to pay for extensive drug safety testing, they are able to sell the drug at a fraction of the cost. If pharmaceutical companies are not able to recoup their developmental and testing costs by selling their name-brand products, they are not going to be able to continue their drug development efforts. In this sense, if pressures by insurance companies to reduce health care costs truly are disincentivizing novel drug development, they may be inadvertently driving up costs by preventing better treatment options from being available in the future.

In the end, however, because the best way to keep costs down is to prevent the need for extensive hospital treatments, like those necessitated by antibiotic-resistant infections, pressure from insurance companies has the potential to be a strong driving force to effect positive change to the health care system. In addition to being a loud voice that can lobby for legislative support to combat antibiotic resistance, health insurance companies also have access to medical and pharmacy claims data, making them well positioned to monitor and provide feedback on antibiotic prescription practices. By working together with health insurers, hospitals, doctors, patients, and regulatory bodies can make data-informed decisions on antibiotic stewardship and other medical practice policies to combat antibiotic resistance.

Lawyers—enforcing accountability fairly. Most people would agree that hospital-acquired infections pose serious safety threats not only to the vulnerable patient population but also to health care professionals, including physicians, nurses, caretakers, and other hospital staff, as well as other visitors who come into physical contact with sick patients. Considering the wide range of bacteria and viruses brought by patients into hospitals and the high volume of people present in hospital settings, it is not surprising that simply going to the hospital is itself a risk factor for acquiring a new infection.

Recognizing the potential dangers of disease spread in the hospital setting, hospitals are required to follow rigid sterilization and treatment protocols to keep the spread of infection under control. Nevertheless, when infections do occur, the hospital may be held liable when the potential risk for infection was not disclosed, when there was a delay in diagnosis or treatment, when improper sanitary procedures were followed, or when equipment was not properly sterilized or discarded. Some lawyers have taken advantage of this situation by charging into the antibiotic resistance picture by specializing in suing doctors and hospitals over hospital-acquired infections, especially those caused by antibiotic-resistant bacteria. So far, most of these cases have been settled out of court as both building and countering a case for medical malpractice take considerable time and resources. Regardless of how justified these cases are, the only way for hospitals to avoid such legal costs is to prevent these infections in the first place. Hospitals are beginning to enforce policies more strictly and are giving physicians and other medical staff the resources and encouragement to clean up their act, literally. Ultimately, regulatory policies and judicial practices will need to be established to ensure a fair balance between acknowledging the inherent risk of hospital-acquired infections despite the best possible efforts to provide a safe environment and demanding accountability when hospitals fall short of meeting the highest standards of health care quality.

On top of the pressure imposed by lawsuits, several state legislatures have enacted laws requiring hospitals to help pay for the cost of hospital-acquired infections, though the specifics of these laws vary from state to state. Some private insurance companies will reimburse hospitals for the cost of treating hospital-acquired infections, but not all are willing to do so. Moreover, a hospital's reputation for having a higher-than-average incidence of difficult-to-treat postsurgical infections can lead to insurance companies declining coverage, while also encouraging lawsuits. The U.S. government's Medicare health insurance program has added further incentives for hospitals to improve their infection prevention and control protocols by supporting the Hospital-Acquired Condition Reduction Program. This program links Medicare payments to hospitals to their annual health care quality performance, which includes hospital-acquired infections metrics.

Governments—global perspectives and considerations. Government plays two vital and defining roles in the global battle against antibiotic resistance. One role is as the financial buttress supporting research, innovation, and the treatment of afflicted individuals. The other role is as the setter and enforcer of regulatory policy. While governments have full control over their geographical areas of influence, pathogens do not respect political borders. In our modern, globalized world, an infectious disease outbreak caused by a difficult-to-treat pathogen occurring in one place is very unlikely to remain confined to that location. As we very forcefully experienced with the COVID-19 pandemic, disease can and will quickly spread to other regions, if not around the world. Consequently, tackling global health issues like the antibiotic crisis requires global cooperation, with financial support from those that can afford it as well as buy-in from poorer countries to properly implement antibiotic stewardship policies.

However, expecting full cooperativity among all governments is probably a bit idealistic. In the real world, infections involving difficult-to-treat pathogens are more likely to emerge in and spread among poor and vulnerable populations, which are more likely to be in more socioeconomically and resource-limited regions of the world. Not only do these countries not have the financial resources to fund the development of their own pharmaceuticals, they also are not always in the best position to properly administer treatments. Added to this challenge is the common practice in many countries of providing over-the-counter antibiotics, i.e., without requiring a prescription from a registered medical practitioner. According to the World Health Organization (WHO), more than 50% of drugs are dispensed and used inappropriately. In developing countries, the rate of self-medication is as high as 90%. This unregulated use of antibiotics can lead not only to other health complications but also to inadvertent spread of antibiotic resistance. A comprehensive and effective solution to the antibiotic crisis will likely require investment, not only in the development of new drugs but also in the health care systems around the world to properly manage and administer them.

ENVISIONING A PATH FORWARD

Overcoming the antibiotic resistance crisis will require a coordinated effort by all stakeholders to support antibiotic discovery and development, to foster appropriate antibiotic management plans, and to formulate new treatment strategies. While balancing the needs and expectations of the various interest groups, there are some actionable steps that can be taken moving forward.

Innovative solutions coming on the horizon from multiple fronts

In chapter 9, we discussed some of the new chemical synthesis strategies for drug discovery and alternative strategies for treatment that are being pursued to fight antibiotic-resistant bacteria. In addition to targeting specific bacteria, there are also broader efforts being made to control the emergence and limit the spread of antibiotic resistance.

Rapid diagnostics at point of care. One of the most important means of sustaining the efficacy of our antibiotics is limiting their use specifically to when they will be most effective. If we can rapidly and accurately identify the pathogen responsible for an infection, physicians can better optimize their selection of which antibiotic to use. More specifically, if the invading pathogen can be identified, treatment can shift from broad-spectrum antibiotics to narrow-spectrum specialized treatments. Likewise, such diagnostic tools can indicate when it is safe and appropriate to discontinue the use of any antibiotic no longer beneficial for clearing a given infection.

Traditional methods for pathogen identification rely on histopathological microscopy and phenotypic microbiological culturing, both of which are limited by the long turnaround time required for getting results, usually ranging from overnight to a few days or even longer. Additionally, some pathogens are difficult to grow in the laboratory, requiring special media or conditions, and some diagnostic tests require specialized personnel to perform or interpret the results. Antibiotic susceptibility testing usually requires up to two more days beyond the initial diagnostic identification, even using advanced modern automated instrumentation systems to facilitate the process.

Nowadays, PCR-based assays are being combined with antibody-based technologies to create rapid diagnostic tests for some of the more frequently encountered pathogens. These rapid diagnostic tests are gradually becoming available for point-of-care use in many clinical settings, e.g., at the patient's bedside, in outpatient clinics, or even in the emergency room. While in practice traditional microbiological laboratory tests have not yet been fully replaced, these newer tests can provide results in as little as a couple hours and are often significantly more sensitive, detecting genes or gene products of characteristic virulence factors (e.g., toxins, adhesins, and capsules). Importantly, many of these newer point-of-care tests can provide information about potential antibiotic resistances, avoiding wasteful ineffective treatments. Moreover, these rapid diagnostic tests are important for developing antimicrobial resistance surveillance capabilities, enabling health care providers to track resistance spread among the population.

In recent years, whole-genome sequencing technologies, combined with bioinformatics tools, have been developed and applied for the detection of all potential pathogens, including any antibiotic resistance genes present, directly from patient samples. While still not practical for rapid point-of-care applications, they have been invaluable for identifying pathogens where all other diagnostic tests have failed. These methods have also contributed significantly to the epidemiological tracking of antibiotic-resistant pathogens both globally and in local hospital or community outbreaks. Some researchers are also turning to intelligent computer algorithms and machine learning approaches in combination with rapid diagnostic tests to distinguish antibiotic-resistant bacteria from sensitive bacteria based on their data profiles. In one such study, machine learning algorithms were used to analyze hundreds of thousands of mass spectrometry profiles from clinically relevant bacterial isolates with known antibiotic susceptibilities to determine which patterns correlated with resistant or sensitive bacteria. The results from a retrospective clinical trial group of patients with *Staphylococcus aureus*, *Escherichia coli*, or *Klebsiella pneumoniae* infections indicated that use of this method would have greatly benefited treatment optimization.

Combination therapies to beat the odds of gaining resistance. As we discussed in chapters 5 and 6, slow-growing bacteria tend to be more resilient to the presence of antibiotics. In some cases, the tolerance is due to the bacteria reducing their growth rate when the drug is around. In other cases, the bacterial population differentiates or splits into an actively growing population that remains susceptible to antibiotics and a dormant population that can persist through the antibiotic treatment. In both cases, even though at first the bacteria are still partially susceptible to the antibiotic, over time the continued exposure to the antibiotic will increase the tolerance level or even lead to the emergence of true resistance.

To prevent the emergence of antibiotic resistance through this tolerance mechanism, physicians have begun developing therapies that include using multiple antibiotics together, taking advantage of the fact that evolution of multiple resistances simultaneously is exceedingly unlikely. Prescribing these combination therapies not only enhances the efficacy of the drug action but also helps prevent new MDR mechanisms from emerging. So far, despite the increased incidence rate of MDR infections, combination therapy is still not commonplace in practice, except in acute, life-threatening infections where the pathogen's identity and its antibiotic susceptibility cannot be determined quickly. Of course, combination therapies would also require the companies that developed the component drugs to agree to manufacture the combination drug together in one formulation for dispensing. While combination therapies are probably not appropriate in all situations, their broader use in treatment regimens should at least be on the table for consideration.

Should we share the burden? Toward a new global business model. Traditionally, funding agencies and private investors have functioned as independent investors with the expectation that they will have full control over any produced research outputs and patents. However, with the advent of globalization, it is no longer reasonable for a single investor or a single government to dictate global health policies, nor is it reasonable for a single government to be financially responsible for funding the world's health care advances. As such, an alternative model to support the antibiotic discovery and development processes has been proposed. By establishing a global collaboration platform in the form of a global fund, private investors, charitable organizations, and government agencies from participating countries can establish funding mechanisms and incentive structures to coordinate efforts by researchers and pharmaceutical companies to take on these global challenges. An example of this type of international financing and partnership organization is The Global Fund to Fight AIDS, Tuberculosis and Malaria that aims to secure investments and resources to end these pandemics. In one proposal, a similar global fund to support combating antibiotic resistance would also buy the sales rights for new antibiotics emerging from this funding and would then manage the supply and distribution of the drugs worldwide. Essentially, this fund would take on global stewardship responsibilities for the new antibiotics, including setting prices for developing countries, establishing guidelines for equitable distribution, and determining where their use would best meet public health needs.

In recent years, nongovernmental organizations and nonprofit private foundations have joined to support these efforts. In 2022, a group of more than 20 international pharmaceutical companies, in collaboration with the WHO, the European

Investment Bank, and the Wellcome Trust, launched one of the largest public-private partnership funds to date, called the AMR Action Fund. The stated goal of this fund was to invest in biotechnology companies willing to commit to bringing two to four new antibiotics or therapeutic treatments currently at the clinical trial stage to market within a decade.

Removing key regulatory barriers

Everyone agrees that it is essential for all new drugs to be thoroughly tested for potential adverse side effects in both animals and humans. However, is it possible that some of our more draconian regulatory policies are unnecessarily obstructive of the drug development process?

How safe is "safe"? Typically, the biggest financial hurdle in the drug development pipeline is the final testing and approval process as the drug goes through clinical trials. It is not surprising then that when drug development costs are discussed, people frequently begin to question how necessary they really are. Is it possible that with more careful analysis, some of the barriers hampering progress of drug candidates through the pipeline could be reduced? For instance, once adequate safety testing in animals has been completed, could some of the extensive challenge and efficacy tests in animals be shortened before moving to human safety testing? Alternatively, if a pharmaceutical company has already tested the drug candidate or similar drug candidates in other countries, could the data from those studies be shared and included in the data used for regulatory approvals in the United States?

When discussing whether antibiotic testing procedures could be streamlined in some way, the tricky question of course is: How safe is "safe"? In practice, the apparent trend lately has been to make the safety and efficacy testing as stringent as possible. In the United States, the Food and Drug Administration (FDA) requires pharmaceutical companies to test enough patients in their final, large clinical trials to be absolutely certain that an antibiotic is safe and effective. The standards set by the FDA in the past have been a model for state-of-the-art safety standards. However, is there a point at which there are diminishing returns to increasing the standards further? Recently, this issue surfaced in the form of an FDA proposal to further decrease the margin of error for determining whether an antibiotic is safe and effective. The bar was already high, but the proposal was to raise the bar even higher. The effect of this increase in the number of subjects tested would be to increase the cost of clinical trials substantially, effectively making antibiotics an even less attractive investment option for pharmaceutical companies.

This is a hard issue for the public to decide. If asked whether we would like our drugs to be even safer and more effective, who among us would say no? But if you were asked whether a small, marginal increase in safety and efficacy is worth stalling the antibiotic discovery pipeline, the answer would probably be different. The problem here for the public is that representatives of the two contending perspectives each have their own secondary motivations, such that neither side can be completely trusted to provide an unbiased perspective. Ideally, some sort of independent auditor could help the public decide what course of action would be most appropriate. At the very least, drug testing data should probably be more easily accessible so analysis by independent parties can be performed.

Another solution some pharmaceutical companies have arrived at to reduce testing costs is to simply perform studies in developing countries. This approach, however, is fraught with potential ethical concerns regarding the exploitation of vulnerable populations and the reliability of trial results. To avoid these ethical issues, the FDA and other regulatory bodies have very strict guidelines on what kind of clinical trial data is acceptable.

Does it have to be "better"? Another line of contention between the pharmaceutical companies and the FDA is the FDA's decision to require that a new antibiotic be demonstrated to be clearly superior to existing antibiotics. This sounds reasonable enough until one realizes that an antibiotic that is equally effective or even slightly less effective than those currently on the market for treating a certain type of infectious disease can suddenly become clearly superior if the bacteria involved become resistant to the antibiotics already on the market.

Vancomycin is a perfect example of the changing antibiotic efficacy landscape. Initially, vancomycin was a less effective option compared to existing antibiotics because it was only effective at targeting Gram-positive bacteria. However, as more and more Gram-positive bacteria became serious MDR threats, suddenly the relative efficacy of vancomycin skyrocketed. The perpetual emergence of antibiotic resistance makes antibiotics a fundamentally different beast compared to other pharmaceuticals that probably should be governed by a separate set of rules.

Beyond the hospital—social responsibility and the fight against antibiotic resistance

The disruption of family life for patients suffering from severe infections is a serious social concern, one that is seldom factored into the cost calculations done by hospital and insurance accountants. Loss of work for the patient, especially for those in lower income brackets, can have compounding effects that magnify the burden placed on other family members. The ensuing financial strain can also interfere with future health insurance coverage, not to mention having other long-term health consequences. But just as there are additional underappreciated hardships that individuals experience once they leave the clinic, there are additional measures that can be taken outside of the hospital setting to abate the antibiotic resistance crisis.

Usually antibiotic-resistant bacterial infections are discussed mainly in the context of hospitals and nursing homes, where crowded populations of highly vulnerable individuals can be found. Unfortunately, MDR bacteria are increasingly becoming a community-wide problem. For example, when the first cases of vancomycin-resistant strains of methicillin-resistant *S. aureus* (MRSA) infection began to appear in the early 2000s, it became clear that these afflicted individuals were not picking up these infections from other patients. Rather, they were acquiring them through contacts in the wider community. These observations are perhaps unsurprising considering that the use of antibiotics in aggregate is actually higher in the community than in hospitals.

At present, in the United States, both hospital-acquired and community-acquired *S. aureus* infections are on the decline overall, and even community-acquired MRSA strains are usually still susceptible to some other antibiotic. The question is how long this will continue to be the case. Clearly, the battle to contain MRSA cannot be waged solely in the easily controlled hospital environment. Rather, efforts need to be taken at the community level. Evidence-based prevention strategies, including rapid diagnosis and on-site antibiotic-susceptibility-guided treatment, will remain a high priority in reducing these types of infections, especially in the community setting. However, additional preventive measures, such as vaccination, hygiene, and other exposure reduction methods, will need to be implemented more widely. Unfortunately, many of these community-acquired infections are disproportionately more prevalent among medically underserved communities, which suffer from additional compounding factors such as poorer education, suboptimal housing conditions, greater food insecurity, patient language barriers, and lack of insurance coverage. Addressing these issues is a prerequisite for an effective overall strategy to control and eventually reduce the rise and spread of antibiotic resistance.

SCIENCE EDUCATION AND DISSEMINATION OF ACCURATE INFORMATION

A massive public campaign to build public awareness of the scale and severity of the antibiotic resistance crisis has been under way for several years now. This campaign has made great strides toward advancing public understanding of the importance of handwashing, clean water, and sanitation, particularly in urban areas, and has already paid dividends in curbing the spread of community-acquired infections. Similar efforts need to be made toward educating policymakers and legislators on the importance of ensuring effective antibiotic stewardship and supporting research and treatment development efforts. Importantly, experts need to be encouraged and rewarded not only for their clinical and research practices but also for their outreach efforts to communicate accurate information to the public both locally and globally.

Science literacy and dissuading science skepticism

Throughout this book we have framed the antibiotic resistance crisis as a global problem that requires big solutions involving multiple different stakeholders at all levels of society. When considering grand problems at such a scale, it can be easy to lose sight of the smaller, more tangible ways that individuals can have a positive impact. One of the most important contributions that an individual can make is to improve their own personal scientific literacy. Far too many people during their education believe that science classes only exist to serve future scientists and doctors. Consequently, when they become adults, they lack the scientific competency to differentiate between scientific studies and conspiracy theories or biased misinformation, thereby creating a general state of confusion. It is no surprise then that there are no unified voices pushing for effective policy changes. Considering the central role that government financial and legislative support plays for funding research and directing health care policy, a population of scientifically literate taxpayers and policymakers is urgently needed to ensure that all aspects of the problem are being addressed adequately and appropriately and that informed decisions are being made.

Thinking outside the box

A key component of a successful path forward is training a cadre of up-and-coming scientists who can think outside the box to solve challenging problems. Support for basic research will be critical not just for understanding how antibiotics work and how bacteria resist them but also for understanding the driving forces behind the spread of antibiotic resistance. Improved understanding of these effects will enable scientists to develop innovative ways of managing MDR bacterial infections. However, coming up with viable creative solutions will require more than just better science. Solutions will also require multidisciplinary understanding of the political, economic, and social factors governing health care advances. True outside-the-box thinking will come from individuals trained at the intersection of these fields who are able to integrate understanding from these different disciplines, while accounting for the different perspectives and motivations of the various stakeholders involved. It is our hope that one of you readers will be such an individual and that you will take inspiration from our discussions to help shepherd us forward into the future.

Index

Revenge of the Microbes: How Bacterial Resistance Is Undermining the Antibiotic Miracle, Second Edition.
Brenda A. Wilson and Brian T. Ho.

B

C

D

E